Sparse Representations for Radar with MATLAB® Examples

Synthesis Lectures on Algorithms and Software in Engineering

Editor
Andreas Spanias, *Arizona State University*

Advances in Modern Blind Signal Separation Algorithms: Theory and Applications
Kostas Kokkinakis and Philipos C. Loizou
2010

Advances in Waveform-Agile Sensing for Tracking
Sandeep Prasad Sira, Antonia Papandreou-Suppappola, and Darryl Morrell
2008

Despeckle Filtering Algorithms and Software for Ultrasound Imaging
Christos P. Loizou and Constantinos S. Pattichis
2008

Sparse Representations for Radar with MATLAB® Examples
Peter Knee

ISBN: 978-3-031-00391-2   paperback
ISBN: 978-3-031-01519-9   ebook

DOI 10.1007/978-3-031-01519-9

A Publication in the Springer series
SYNTHESIS LECTURES ON ALGORITHMS AND SOFTWARE IN ENGINEERING

Lecture #10
Series Editor: Andreas Spanias, *Arizona State University*
Series ISSN
Synthesis Lectures on Algorithms and Software in Engineering
Print 1938-1727   Electronic 1938-1735

Sparse Representations for Radar with MATLAB® Examples

Peter Knee

Sandia National Laboratories, Albuquerque, New Mexico

SYNTHESIS LECTURES ON ALGORITHMS AND SOFTWARE IN ENGINEERING #10

ABSTRACT

Although the field of sparse representations is relatively new, research activities in academic and industrial research labs are already producing encouraging results. The sparse signal or parameter model motivated several researchers and practitioners to explore high complexity/wide bandwidth applications such as Digital TV, MRI processing, and certain defense applications. The potential signal processing advancements in this area may influence radar technologies. This book presents the basic mathematical concepts along with a number of useful MATLAB® examples to emphasize the practical implementations both inside and outside the radar field.

KEYWORDS

radar, sparse representations, compressive sensing, MATLAB®

Contents

List of Symbols

The following definitions and relations between symbols are used throughout this text unless specifically noted.

$*$	Convolution operator
$x(t)$	Continuous variable
$x[t]$	Discrete variable
$\boldsymbol{x}$	Vector variable
$\hat{\boldsymbol{x}}$	Estimate for vector variable
$\hat{\boldsymbol{x}}^n$	Estimate for vector variable at iteration
X	Matrix variable
$X(i, j)$	Element from row, column of matrix
$\boldsymbol{X}^{\dagger}$	Pseudo-inverse of matrix
$\mathbb{R}$	Set of real numbers
$\text{Re}\{x\}$	Real portion of complex variable
$\mathbb{R}^{n \times m}$	Euclidian space for real matrices of size
$\emptyset$	Null set
$\underline{\boldsymbol{0}}$	Vector of zeros
$\boldsymbol{1}$	Vector of ones
$(\cdot)^T$	Transposition
$\boldsymbol{I}$	Identity matrix
$\|\cdot\|_n$	l_n-vector norm
P_D	Probability of detection
P_{FA}	Probability of false alarm
P_M	Probability of miss
P_{CC}	Probability of correct classification
ΔCR	Cross range resolution
ΔR	Range resolution
T_p	Pulse repetition period
τ_b	Pulse width
λ	Wavelength
B	Bandwidth

List of Acronyms

The following acronyms are used throughout this text.

ADC	Analalog-to-Digital Converter
ATR	Automatic Target Recognition
CFAR	Constant False Alarm Rate
CW	Continuous Wave
DSP	Digital Signal Processing
DTM	Digital Terrain Map
EM	Expectation Maximization
FFT	Fast Fourier Transform
FT	Fourier Transform
IFSAR	Interferometric Synthetic Aperture Radar
IRE	Instititue of Radio Engineers
ISOMAP	Isometric Mapping
LDA	Linear Discriminant Analysis
LLC	Local Linear Coordination
LLE	Local Linear Embedding
LP	Linear Programming
LTSA	Local Tangent Space Alignment
MDL	Minimum Description Length
MDS	Multi-Dimensional Scaling
MIMO	Multi-Input Multi-Output
MP	Matching Pursuit
MTI	Moving Target Indication
MVU	Maximum Variance Unfolding
OMP	Orthogonal Matching Pursuit
PCA	Principal Component Analysis
PPI	Plan-Position Indicator
RP	Random Projection
SAR	Synthetic Aperture Radar
SFR	Step Frequency Radar
SIFT	Scale-Invariant Feature Transform
SNR	Signal-to-Noise Ratio
TV	Total Variation

Acknowledgments

Portions of this work were supported by Raytheon Missile Systems (Tucson) through the NSF ASU SenSIP/Net-Centric Center and I/UCRC. Special thanks to Jayaraman Thiagarajan, Karthikeyan Ramamurthy, and Tom Taylor of Arizona State University, as well as Visar Berisha, Nitesh Shah and Al Coit of Raytheon Missile Systems for their support of this research.

Peter Knee
September 2012

CHAPTER 1

Radar Systems: A Signal Processing Perspective

In the late 1800's Heinrich Hertz demonstrated a remarkable phenomenon; radio waves are deflected and refracted in much the same manner as light waves [1]. This demonstration proved the validity of Maxwell's electromagnetic theory and laid the groundwork for the development of the modern radar systems. In fact, the term "radar" has become so predominant in our vernacular that its original development as an acronym for "*ra*dio *d*etection *a*nd *r*anging" has given way to its use as a standard English noun [2].

Advances in radar systems and the identification of new applications are driven by enhancements in digital signal processing algorithms. Consider the basic block diagram for a conventional pulsed radar system with a superheterodyne receiver in Figure 1.1. Analog-to-digital (A/D) converters that must sample on the order of tens or hundreds of megahertz and the large amounts of data that must be processed necessitate the need for advanced signal processing algorithms. For this reason, radar engineers pay particular attention to advancements in the field of digital signal and image processing. Typically, DSP techniques have concentrated on improving a single aspect of a radar system such as range resolution or Doppler estimation. The objective of this book is to highlight the potential of sparse representations in radar signal processing.

Early work in sparse representations by mathematicians and engineers concentrated on the potential for finding approximately sparse solutions to underdetermined systems of linear equations. The recent guarantees for exact solutions computed using convex optimization techniques continue to generate interest within academia. Sparse representations have several applications across several disciplines [3].

1.1 HISTORY OF RADAR

The demonstration of the reflection of radio waves by Hertz in 1886 paved the way for the German engineer Christian Hülsmeyer [4] to develop the first monostatic pulsed radar for collision avoidance in maritime vessels. Surprisingly, radar systems did not receive attention again until 1922 when S.G. Marconi [5] emphasized its importance to the Institute of Radio Engineers (IRE). Independently, work also began at that time at the U.S. Naval Research Laboratory on a system that utilized widely separated transmitters and receivers for ship detection. Nearly ten years later, this work received a patent as the first bistatic (separate receiver and transmitter) continuous wave (CW) radar system.

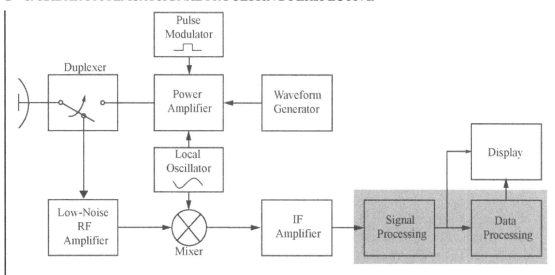

Figure 1.1: Block diagram of a conventional pulsed monostatic radar, with the potential impacts of sparse representations highlighted in gray.

Numerous other reports of ship detections using similar systems occurred throughout the 1920s, however this early technology never saw widespread commercial utilization.

The bistatic CW radar was actually a result of the accidental detection of an aircraft as it passed between the transmitter and receiver of a radio system [2]. This setup was burdensome and required the target to pass between the transmitter and receiver of a radio system. The emergence of long-range, high-altitude military aircraft after World War I ignited the development of colocated transmitter and receiver radar systems utilizing pulsed waveforms. Work by A. Hoyt Taylor and Leo C. Young began in the U.S. at the U.S. Naval Research Laboratory in 1934 on single site, pulsed radar systems [4]. Similar systems, operating in the 100–200 MHz range, were developed throughout the world. It was not until after the end of World War II, with the release of the microwave power tube, that radar systems began to operate at the high-end microwave frequencies used today.

The end of the war did not signal the end of development. In addition to continuous enhancements for military radar capabilities, modern commercial systems continue to see expansive development. Signal processing techniques including pulse compression, synthetic aperture radar imaging, and phased array antennas have provided tremendous capabilities both in the commercial and military domains. The advancements across the radar industry continue to this day, with military applications still driving research and development. Going forward, advances in digital signal processing technology will be required for the future state-of-the-art radar systems.

1.2   CURRENT RADAR APPLICATIONS

Most scientists and practitioners are well aware of the capabilities of radar systems to *detect* and *track* targets, likely stemming from their familiarity with displays such as the plan-position indicator (PPI) display shown in Figure 1.2. The 2-D display shows scatterer reflection magnitudes with range and azimuth displayed in polar coordinates. These displays, along with its numerous variants, were vital in presenting accurate information to a user for manual extraction of target speed and direction. The development of enhanced digital signal and data processing techniques has allowed for the automatic extraction of this information, relegating the user to a bystander. Perhaps a less-known function of radar however is that of target scene *imaging*. The development of both 2-D and 3-D images of an area allows for analysis for a variety of purposes, including surveillance and intelligence, topological mapping, and Earth resource monitoring. The benefit of radar imaging lies in its ability to operate effectively in adverse weather conditions, including rain, fog and cloud cover. This capability is a result of the use of signal frequencies that reduce atmospheric attenuation of the transmitted waves. Additionally, radar imaging systems have the capability to operate at night when optical imaging systems cannot operate at all.

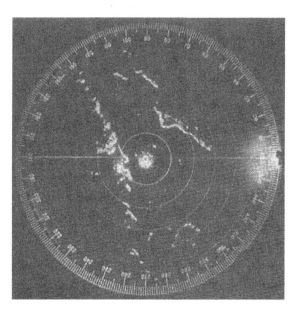

Figure 1.2: Example of a PPI (plan position indicator) display. (From Middleton and Mair, *Radar Technical Overview.* copyright © National Research Council of Canada, http://www.ieee.ca/ millennium/radar/radar_technical.html.)

Some of the more popular radar technologies include [4]:

• meteorological radars;

- military surveillance, including target detection, recognition and tracking;

- interferometric SAR, for 3-D scene images;

- ground-penetrating radar;

- ballistic missile defense;

- air-traffic control;

- environmental remote sensing;

- law-enforcement and highway safety; and

- planetary exploration.

Figure 1.3 gives a slight indication of the diversity in commercial and military radar systems. Many readers may be familiar with the systems found at local airports but these images show the vast differences in currently available radar applications.

1.3   BASIC ORGANIZATION

This book will present a practical approach to the inclusion of the sparse representation framework into the field of radar. The relevant mathematical framework for the theory of sparse representations will be discussed but emphasis will be placed on providing examples using the included MATLAB® scripts. The ability to immediately manipulate operating parameters is intended not only to facilitate learning but also provide the opportunity to quickly integrate emerging signal processing techniques into radar applications.

Chapter 2 presents the basics in sparse representation theory. Emphasis is again not focused on the mathematics but on the application of the theory. Simple signal processing examples are included to demonstrate the utility of the sparse representation framework. A discussion on reduced representations for radar would not be complete without at least a tertiary discussion on the assumption of sparse representations for radar signals. To this end, Chapter 3 presents a brief review on common dimensionality reduction techniques. Examples of reduced dimensional representations for radar, specifically in the imaging domain, are included. The final two chapters of this text serve to introduce radar signal processing fundamentals and the radar applications in which sparse representations have already been explored. We also take the opportunity to present our current work in the area of automatic target recognition (ATR).

It should be noted that *compressed sensing* is a recent branch of sparse representation theory that is quickly attracting a lot of attention. The exploitation of signal sparsity allows for sampling that is very effective. As the work on this topic alone is immense, the theory will not be considered in this book. However, seeing as it is a branch of sparse representations, the applications in the field of radar will be presented to illustrate the effect sparse representations have had on radar technologies.

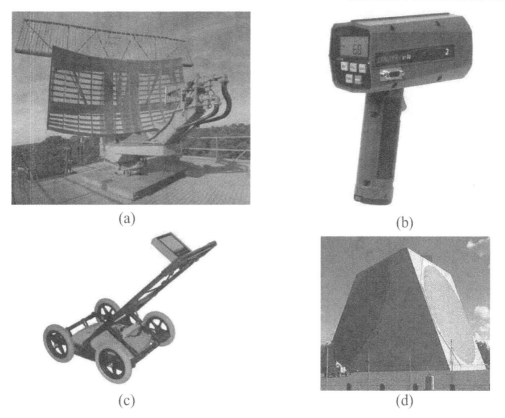

Figure 1.3: (a) airport surveillance radar for vicinity surrounding airport, from the Federal Aviation Administration, `http://en.wikipedia.org/wiki/File:ASR-9_Radar_Antenna.jpg`; (b) speed monitoring radar gun, copyright © Applided Concepts, Inc. `http://www.stalkerradar.com/law_basic.shtml`; (c) ground-penetrating radar used to locate buried cables, pipes, power lines, or other objects, copyright © SPX Corporation, `http://www.radiodetection.com/`; and (d) phased array radar that allows for scanning without radar motion, from the U.S. Army Corps of Engineers, `http://en.wikipedia.org/wiki/Radar`.

As the two are intertwined, we hope that the use of *sparse representation* and *compressed sensing* interchangeably does not cause any confusion.

CHAPTER 2

Introduction to Sparse Representations

Be it physical or theoretical, solutions to complex problems can often be determined by following the age-old mantra: "Work smart, not hard." Scientifically, this principle is referred to as Occam's razor, or the law of parsimony. This guiding principle generally recommends selecting the hypothesis that makes the fewest new assumptions when two competing hypotheses are equivalent in all other regards. An interesting example would be the consideration of the orbit of the planets; do they revolve around the Earth or around the sun? Both are possible but the explanation for the planets revolving around the sun requires far less complexity and is thus preferred.

For decision-making tasks, the application of Occam's razor has led to the development of the principle of minimum description length (MDL) [6]. MDL is used as a heuristic, or general rule, to guide designers towards the model that provides the most concise system representation. For high-dimensional signals or systems, this implies that small or even single subsets of features that accurately describe their complexity are preferred. The process of extracting meaningful lower-dimensional data from high-dimensional representations occurs naturally within the human visual system. In a single instant, the human brain is confronted with 10^6 optic nerve fiber inputs, from which it must extract a meaningful lower-dimensional representation [7]. Theoretically, scientists encounter this same problem for high-dimensional data sets such as 2-D imagery. Numerous compression algorithms, such as JPEG 2000 image coding standard, utilize wavelet coefficients to extract salient features from the high-dimensional data, negating the need to retain all the image pixels. Image storage and transmission are easier using lower-dimensional representations. Across the application spectrum, the ability to develop a low-dimensional understanding of high-dimensional data sets can drastically improve system performance.

Throughout the signal processing community, there has been large-scale interest in estimating lower-dimensional signal representations with respect to a dictionary of base signal elements. When considering an overcomplete dictionary, the process of computing sparse linear representations has seen a surge of interest, particularly in the areas of classification [8] and signal acquisition and reconstruction [9, 10]. This interest continues to grow based on recent results reported in the sparse representation literature: If a signal is sufficiently sparse, the optimal signal representation can be formulated and solved as a general convex optimization problem [10]. The specifics of these two technical arguments will be discussed in more detail in Section 2.1.

As mentioned, compressive sensing is a research area associated with sparse representations tailored to signal acquisition and multimodal sensing applications. Compressive sensing [11] was designed to facilitate the process of transform coding, a typically lossy data compression technique. The common procedure, typical in applications such as JPEG compression and MPEG encoding, is to: (1) acquire the desired N-length signal x at or above the Nyquist sampling rate; (2) calculate the transform coefficients; and (3) retain and encode the K largest or the K most perceptually meaningful coefficients and their locations. By directly acquiring a compressed signal representation, which essentially bypasses steps (1) and (2), compressive sensing has been shown to be capable of significantly reducing the signal acquisition overhead. More importantly, the theoretical developments from the area of sparse representations have provided the ability to accurately reconstruct the signal using basic optimization processes. Compressive sensing without the ability to reconstruct the signal is merely a means of dimensionality reduction.

While early signs of sparse representation theory appeared in the early 1990s, this field is still relatively young. This chapter presents the basic mathematics along with intuitive and simple examples in an effort to provide the background needed to understand the utility of sparse representations. As such, the core concepts presented in this chapter provide the basics of signal sparsity required to cover the radar topics presented later in this book. For more detailed discussions on both the theory and applications of sparse representations, we encourage the reader to consult the numerous references provided throughout the chapter.

2.1   SIGNAL CODING USING SPARSE REPRESENTATIONS

The basic aim in finding a sparse representation for a signal is to determine a linear combination of elementary elements that are able to adequately (according to some metric) represent the signal. Consider a set of unit-norm column vector elements, $[d_1, \ldots, d_N]$, stacked into a matrix $D \in \mathbb{R}^{M \times N}$, known as an N-element dictionary. The linear combination of all elements in the dictionary can be written as

$$y = x_1 d_1 + \ldots + x_N d_N , \qquad (2.1)$$

where x_n are scalar coefficients. In matrix notation this is equivalent to

$$y = D x_0 \in \mathbb{R}^M , \qquad (2.2)$$

where x_0 is a coefficient vector whose entries are the scalar coefficients of (2.1). A sparse representation for the signal y indicates that the number of non-zero coefficients in the representation vector x_0 is less than M. Typical situations result in the percentage of non-zero coefficients being between 0 and 30% with algorithm breakdowns occurring with up to 70% non-zeros [8].

For $M > N$, the system of equations described by $y = D x_0$ contains more equations than unknowns. When a non-trivial solution does exist for this *overdetermined* system of equations, the solution can be approximated using the method of least squares [12]. Referred to also as the *method of frames* [13], this is equivalent to finding a solution that minimizes the Euclidian distance between the true and reconstructed signals, i.e.,

$$\hat{x}_0 = \min_{x_0} \|y - Dx_0\|_2^2 . \tag{2.3}$$

The approximation for x_0 can be found using the pseudo-inverse of the $D^\dagger = (D^T D)^{-1} D^T$ dictionary matrix, i.e., $x_0 = D^\dagger y$. More often, however, we are concerned with the case where we have more unknowns than equations (i.e., $M < N$). The solution to (2.3) typically contains an infinite number of solutions. Using the Euclidian norm as the representation metric, the minimum norm solution can be found by minimizing the length of the coefficient vector, i.e.,

$$\hat{x}_0 = \min_{x_0} \|x_0\|_2 \text{ subject to } y = Dx_0 . \tag{2.4}$$

The pseudo-inverse $D^\dagger = D^T (DD^T)^{-1}$ again provides the minimum norm solution. Unfortunately, whether the system is over or underdetermined, the solution $\hat{x}_0$ is typically not informative, especially in the sparse representation framework, as it contains a large number of non-zero elements. The relatively large percentage of non-zero elements is a result of the minimum energy constraint, which tends to prefer numerous smaller elements to a few larger elements. For a geometric interpretation of this fact, see Section 2.2.

Instead, we can explicitly seek a sparse solution to $y = Dx_0$ by formulating the linear system as an l_0-norm minimization problem

$$\hat{x}_0 = \min_{x_0} \|x_0\|_0 \text{ subject to } y = Dx_0 , \tag{2.5}$$

where $\|\cdot\|_0$ is referred to as the l_0-norm. The coefficient vector $\hat{x}_0$ that contains the fewest number of non-zero elements, also known as the sparsest vector, is now the preferred solution. Unfortunately, this *minimum weight solution* [14] is NP-hard and has been shown to be difficult to even approximate [15].

Despite the difficulty in computing the minimum weight solution, sparse representations continue to generate a lot of interest. The pursuit of a tractable, minimum weight solution has been made possible due to one important result: if the solution is sparse enough, the solution to the l_0-minimization of (2.5) can be unique and equal to the solution of the convex l_1-minimization problem [10]

$$\hat{x}_1 = \min_{x_0} \|x_0\|_1 \text{ subject to } y = Dx_0 , \tag{2.6}$$

where $\|\cdot\|$ is the l_1-norm given by the sum of the magnitudes of all vector elements. Various approaches exist for finding solutions to (2.6) and will be the subject of Section 2.3. The formulation of the sparse representation problem in (2.6) as a convex optimization problem allows for the efficient (polynomial time) calculation of the solution using interior-point programming methods [16]. Additionally, emerging greedy pursuit methods have shown to be extremely versatile in providing approximate solutions to the non-convex l_0-minimization problem.

One noticeable assumption at this point is that the signal y can be expressed exactly as a sparse superposition of the dictionary elements. In practice, noise may make this a strong assumption although it can be accounted for by assuming that $y = Dx_0 + z$, where $z \in \mathbb{R}^M$ is small, possibly dense noise. The l_1-minimization problem of (2.6) becomes a convex optimization problem of the form

$$\hat{x}_1 = \min_{x_0} \|x_0\|_1 \text{ subject to } \|Dx_0 - y\| \leq \epsilon \; . \tag{2.7}$$

2.2   GEOMETRIC INTERPRETATION

The distributive nature of the l_2-norm in developing a sparse representation is important and will be addressed here. A simple geometric interpretation will clearly show that as we move from l_2 regularization towards l_0, we promote sparser and more informative solutions. We will consider l_p-"norms" for $p < 1$ although the formal "norm" is not defined since the triangle inequality is no longer satisfied.

To see this "promotion" of sparsity, consider the following generic problem:

$$\min_{x} \|x\|_p^p \text{ subject to } y = Dx \; . \tag{2.8}$$

The solution set for an underdetermined linear system of equations $y = Dx \in \mathbb{R}^n$ is a subspace of $\mathbb{R}^n$. If we consider a particular solution, x_0, the feasible set is a linear combination of x_0 and any vector from the null-space of A. Geometrically, this solution set appears as a hyperplane of dimension $\mathbb{R}^{n-m}$ embedded in $\mathbb{R}^n$ space [3].

```
%lpNorm.m
x = -1.25:.01:1.25;
m = -.66; b = 1;
yl = m*x+b;
p = [.4 .9 1 1.5 2];

figure;
for ii = 1:length(p)
    [Np ind] = min((abs(yl).^p(ii)+abs(x).^p(ii)).^(1/p(ii)));
    xp = linspace(-Np,Np,1000);
    yp = (Np^p(ii)-abs(xp).^p(ii)).^(1/p(ii));

    subplot(1,length(p),ii);
    plot(xp,yp,'-k',xp,-yp,'-k');hold on;
    plot(x,yl,'-r'); xlim([-1.25 1.25]);ylim([-1.5 1.5]);axis square;
    title(sprintf('P = %.1f',p(ii)));xlabel('x_1');ylabel('x_2');
end
```

Program 2.1: Computation of l_p-norm balls.

For the purposes of illustration, we will consider a 2-D example, in which $D \in \mathbb{R}^{2 \times 3}$ so that the solution set is a line in the two-dimensional subspace. The solution to (2.8) is then found by "blowing" an l_p-ball centered around the origin until it touches the hyperplane solution set. This methodology is demonstrated using script Program 2.1. Results for values ranging from $p = .4$ to $p = 2$ are shown in Figure 2.1. The norm-ball for $p \leq 1$ contains sharp corners whose points lie on the coordinate axes. It is the solutions at the corners of the norm-ball where 1 or more coefficients are zero that promote sparsity in a representation. Conversely, the intersection of the solution set with the l_p-norm balls for $p > 1$ occurs off a cardinal axis. The solution thus contains all non-zero coefficient values. Stated more explicitly, the intersection of the solution subspace with a norm ball of $p \leq 1$ is expected to occur on the cardinal axes, forcing coefficient values to be zero in the sparse solution.

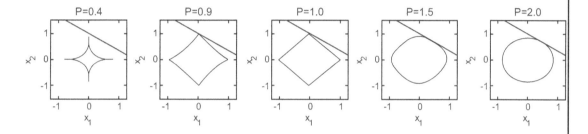

Figure 2.1: The intersection between the l_p-ball and the feasible solution set for the linear system $y = Dx$. Values of $p \leq 1$ force a sparse solution that lie on the "corners" of the respective norm-balls, whereas the solution for larger values is a non-sparse solution resulting in the most feasible point closest to the origin.

2.3   SPARSE RECOVERY ALGORITHMS

Theoretically, it has been noted that it is possible, under certain circumstances, to recover the solution to the l_0-minimization problem using the convex l_1-formulation. The burden now falls on algorithmic designs for the solution of the l_1-minimization problem in (2.6). Under the compressive sensing framework, this implies that it is possible to design non-adaptive measurements that contain the information necessary to reconstruct virtually any signal. The recent advances in sparse representations and compressive sensing have driven rapid development of sparse reconstruction algorithms. The consideration of all the approaches is beyond the scope of this book, however, the natural discrimination of the approaches into two basic methods allows for a brief introduction that can provide the background needed for the examples given in the subsequent section.

2.3.1 CONVEX OPTIMIZATION

The highly-discontinuous l_0-norm makes the computation of the ideal sparse solution often not tractable. Numerical approximations require that the l_0-norm be replaced by a continuous or even smooth approximation. Examples of such functions include the replacement with l_p-norms for $p \in (0, 1]$ or smooth functions such as $\sum_j \log(1 + \alpha x_j^2)$ or $\sum_j (1 + \exp(-\alpha x_j^2))$. The l_1-norm problem specifically has been addressed using basic convex optimization techniques including conversion to a linear programming problem that can be solved using modern interior point methods, simplex methods, homotopy methods, or other techniques [3].

Recall the l_1-optimization problem

$$\hat{x}_1 = \min_{x_0} \|x_0\|_1 \text{ subject to } \|Dx_0 - y\| \le \in . \tag{2.9}$$

Suppose the unknown x_0 is replaced by $x_0 = u - v$ where $u, v \in \mathbb{R}^n$ are both non-negative vectors such that u takes all the positive entries in x_0 and v takes all the negative entries. Since we can write that $\|x_0\|_1 = 1^T x_0 = 1^T (u - v) = 1^T z$ where 1 is a vector of ones and $z = [u^T, v^T]^T$, the l_1-optimization problem can be written as

$$\hat{x}_1 = \min_{x_0} 1^T z \text{ subject to } \|[D, -D]z - y\| \le \in \text{ and } z \ge 0 \tag{2.10}$$

which now has the classical form for a linear-programming (LP) problem. Simplex algorithms like the Dantzig selector [17] solve LP problems by walking along a path on the edge of the polytope solution region until an optimum is reached. Interior point methods, albeit slower, solve the problem in polynomial time [16] and are another option when considering approximations for (2.10).

2.3.2 GREEDY APPROACH

Greedy algorithms are iterative approaches that seek to locate the dictionary atoms that best approximate the properties of a given signal. As the name overtly implies, the algorithm is designed to be *greedy* at each stage by selecting the best possible approximation at the time, hoping to achieve a satisfactory final solution in the end. There are a number of greedy methods, however, the most popular continue to be the basic matching pursuit (MP) [18] and orthogonal matching pursuit (OMP) [19] algorithms. MP continues to see throughput increases with very efficient $\mathcal{O}(N \log M)$ (per iteration) implementations that have led to practical large scale application usage [20]. Current iterations of the OMP algorithm are much more computationally intensive, however, the algorithm has shown superior approximation performance.

Consider the iterative approximation of the signal y using the dictionary D, where each column vector d_i will be referred to as an atom. The greedy algorithms approximate the signal in iteration n as

$$\hat{y}^n = D_{\Gamma^n} x_{\Gamma^n} , \tag{2.11}$$

where the vector x is the sparse representation for y and Γ^n denotes the set containing the indices of the atoms selected up to and including iteration n. The approximation error or residual is calculated as

$$r^n = y - \hat{y}^n \, . \tag{2.12}$$

The residual at each iteration is used to determine the new atoms that will be selected in subsequent iterations to achieve a better approximation. While the selection strategy for the new atom does differ slightly across greedy algorithms, MP and OMP choose the index for the selected atom at iteration n as

$$i^n = \arg\max_i |D^T r^{n-1}| \, . \tag{2.13}$$

MP and OMP fall into a class of greedy algorithms referred to as *directional updates* [20], all of which share a common algorithm structure, differing only in the method in which the update direction at each iteration is computed. As given in [20], the algorithm can be summarized as follows.

Initialize
$$r^0 = y$$
$$x^0 = \underline{0}$$
$$\Gamma^0 = \emptyset$$
For n = 1; n:=n+1
1) $g^n = \Phi^T r^{n-1}$
2) $i^n = \arg\max_i |g_i^n|$
3) $\Gamma^n = \Gamma^{n-1} \cup i^n$
4) Calculate update direction p_{Γ^n}
5) $c^n = d_{\Gamma^n} p_{\Gamma^n}$
6) $a^n = \dfrac{\langle r^n, c^n \rangle}{\|c^n\|_2^2}$
7) $x_{\Gamma^n}^n := x_{\Gamma^n}^{n-1} + a^n p_{\Gamma^n}$
8) $r^n = r^{n-1} - a^n c^n$
Break if $\|r^n\|_2 \le \epsilon$
Output r^n, x^n

Algorithm 2.2: Greedy directional update algorithm.

As mentioned, the choice of the directional update at step (4) in the iterative loop determines the type of greedy algorithm. The original matching pursuit algorithm chose a directional update equivalent to $p_{\Gamma^n} = \epsilon_{i^n}$ where ϵ_k is the Dirac basis $\mathbb{R}^N$ [20]. This process is a special case of a technique called projection pursuit, common throughout the statistics literature [21]. The asymptotic convergence is based on the orthogonality of the residual to the previously selected dictionary atom. Orthogonal matching pursuit improves performance by guaranteeing projection onto the span of the dictionary elements in no more than N steps by ensuring full backward orthogonality of the error at each iteration. The directional update for the OMP algorithm is achieved by setting $p_{\Gamma^n} = D^{\dagger} r^{n-1}$.

```
function [A]=OMPnorm(D,X,L,eps)
%OMPNORM            Sparse coding of a group of signals based on a
%                   dictionary and specified number of atoms
%
% [USAGE]
%    A = OMPnorm(D,X,L,DEBUG)
%
% [INPUTS]
%    D       M x K overcomplete dictionary
%    X       Data vector or matrix.  Must be of size M x P, where P is
%            number of signals to code
%    L       Maximum number of coefficients for each signal
%    eps     Stopping criterion (stop when l2 norm of residual is less
%            than eps)
%
% [OUTPUTS]
%    A       Sparse coefficient matrix of size K x P
%

P=size(X,2);
[M K]=size(D);

%Ensure that the dictionary elements have been normalized
Dnorm = D./repmat(sqrt(sum(D.^2)),[M 1]);
A = zeros(K,P);
for k=1:1:P,
    x             = X(:,k);
    residual      = x;
    indx          = zeros(L,1);
    resids        = zeros(K+1,1);
    for j = 1:1:L
        proj            = Dnorm'*residual;
        [maxVal,pos]    = max(abs(proj));
        pos             = pos(1);
        indx(j)         = pos;
        a               = pinv(D(:,indx(1:j)))*x;
        residual        = x-D(:,indx(1:j))*a;
        resids(j+1)     = sum(residual.^2);

        %Break if error has reach sufficient value
        if sum(residual.^2) < eps
            break;
        end
    end;
    A(indx(1:j),k)=a;
end
```

Program 2.3: Orthogonal matching pursuit function.

The simplicity of the MATLAB$^{\circledR}$ implementation for the OMP algorithm is shown in script Program 2.3. Unfortunately, while implementation is simple, computational complexity can be quite high for the OMP algorithm. A recent study [22] detailing the processing and memory requirements for the MP and OMP algorithms shows the dramatic increase in required resources for the OMP algorithm. The improvement in approximation estimation is only seen when the ratio of non-zero elements to the number of observations increases above 20%. Regardless, both algorithms have been shown extensively to provide adequate sparse approximations while significantly reducing the theoretical performance gap that exists between the greedy approaches and their linear programming counterparts [23].

2.4 EXAMPLES

This section is devoted to demonstrating that the theory of sparse representations and the development of increasingly efficient and accurate l_1-minimization solvers could have ramifications in nearly every aspect of everyday life. We present two examples: one detailing the development of a non-uniform sampling theorem [24] for a 1-D signal that reduces the required sampling rate, and the other, an image reconstruction technique that has proven to be advantageous for fields ranging from medicine to radar.

2.4.1 NON-UNIFORM SAMPLING

Consider a generic 1-D analog signal $x(t)$ and its corresponding discrete-time representation $x[n]$. N samples of the analog signal are computed by setting $t = nT_s$, where $n = 0, \ldots, N - 1$ and T_s is known as the sampling period, or the spatial time in between samples. Shorthand notation throughout the book will omit the discrete-time index n and emphasize the 1-D nature of the discrete signal by using the vector notation $\boldsymbol{x}$. The development of the compressive sensing framework arose due to the desire to circumvent the well-known Shannon-Nyquist sampling criterion, an increasingly-limiting design requirement as signal processing demands continue to grow. The theorem states that a signal strictly bandlimited to a bandwidth of B rad/s can be uniquely represented using sampled values that are spaced at uniform intervals no more than π / B seconds apart. Noting that the sampling period $T_s = 1/f_s$, where f_s is referred to the ordinary sampling frequency (measured in Hertz) and is related to the angular frequency ω (measured in radians per second) by $\omega = 2\pi f$, the sampling criterion requires that the signal be sampled at a rate greater than twice the maximum frequency component of the signal ($\omega_{\max} = B \rightarrow f_s > \frac{B}{\pi} = \frac{2\pi f_{\max}}{\pi} = 2 f_{\max}$).

The sum of three sinusoids at varying frequencies will be used to display the utility of the sparse representation framework. Our goal is to demonstrate the ability to sample at a rate much less than $2 f_{\max}$ while retaining the ability to accurately reconstruct the signal. Consider the discrete-time signal:

$$\boldsymbol{x} = x[n] = \sin(2\pi 1000 n T_s) + \sin(2\pi 2000 n T_s) + \sin(2\pi 4000 n T_s) . \qquad (2.14)$$

As shown in the upper-left plot of Figure 2.2, the signal is clearly not sparse in the time domain. Recalling the definition of the Fourier transform (FT), $X(\omega) = \int_{-\infty}^{\infty} x(t)e^{-j\omega t}$, and the linearity of the transform, it is easily verifiable that the frequency domain signal consists of 6 sharp peaks at $\pm$ the principal frequencies of each sinusoid. More importantly, the sparse signal f can be obtained by projecting the signal onto a set of sinusoidal basis functions stacked row-wise into a linear transformation matrix Φ (i.e., $f = \Phi x$). The matrix Φ is known as the discrete Fourier transform (DFT) matrix and can be generated in MATLAB® using either the `dftmtx` function or `fft(eye(N))`. To capture the underlying properties of the signal, we must sample at M non-uniform locations. Randomly sampling the signal x is equivalent to multiplication by a sampling matrix $S \in \mathbb{R}^{M \times N}$, where each row of the matrix contains a single unit element in the column corresponding to the ith sample of the signal x to be extracted. In keeping with the compressive sensing convention, we thus have

$$\hat{x} = \frac{1}{N} S \Phi^T f , \tag{2.15}$$

where the $(\cdot)^T$ refers to the matrix transposition. The reconstructed signal $\hat{x}$ can be determined by finding the sparse representation for f according to the l_1-minimization technique presented in (2.6).

For the purposes of this example, we have sampled the signal x in (2.14) at a rate of 10 kHz for a duration of .1 s ($N = 1028$ samples). It is assumed that a large number of non-zero elements exist in the sparse coefficient vector f, so it is expected that many of the coefficients returned by the OMP algorithm will be very near zero. Additionally, we have randomly selected approximately 15% of the available signal samples for reconstruction. This is in accordance with the requirements for the OMP algorithm described in [20]. The script used to generate the results is included in Appendix A.1.

Figure 2.2 details the results with the time domain signals in the left column and the frequency domain signals (coefficients) on the right. The figure at the top left is the fully sampled N-length signal with the randomly selected M samples shown using blue circles. The frequency domain signal verifies the presence of the three sinusoids. The second row of figures indicates the time-domain reconstruction results (left) using the frequency domain coefficients (right) computed using the minimum norm solution for our underdetermined linear system ($\hat{f} = (S\Phi^T)^{\dagger}x = \Phi S^T (SS^T)^{-1}x$). The distribution of energy across all the coefficients using this *naïve* solution is evident. The third row shows the performance of the sparse representation framework in reconstructing the signal and estimating the Fourier coefficients. The reconstruction is not exact (due to the estimation of $M/4$ non-zero coefficient values) but the iterative nature of the OMP algorithm combined with the reduced sampling rate in generating an accurate reconstruction are very promising.

It is important to emphasize the utility of this approach. Rather than acquiring all samples of the signal, transforming and coding the coefficients for storage, a reduced number of random samples can be extracted and stored directly. The nearly 85% reduction in the Nyquist sampling rate will prove to be very advantageous in a society focused on technological improvements. For further

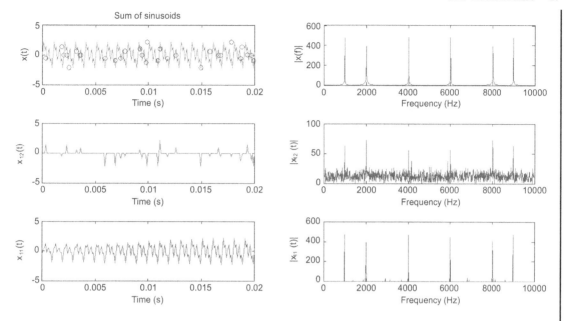

Figure 2.2: Top: (left) original time-domain signals with random samples in blue and (right) frequency domain coefficients. Middle: (left) Naïve reconstruction solution with time-domain signal and (right) transform coefficients on the right. Bottom: (left) l_1-minimization reconstruction computed OMP norm and (right) Fourier coefficients of reconstructed signal. All plots were generated using the script in Appendix A.1.

implementation possibilities, we urge the reader to consider using the l_1-MAGIC package [25] to compute the sparse Fourier solutions using the basis pursuit methods described in Section 2.3.1. An alternative and very useful example is presented using the discrete cosine transform [26].

2.4.2 IMAGE RECONSTRUCTION FROM FOURIER SAMPLING

The utility of the sparse representation framework and the improvements to existing signal processing algorithms becomes evident when we consider the classical tomography problem: reconstruct a 2-D image $f(t_1, t_2)$ from samples $\hat{f}|_\Omega$ of its discrete Fourier transform on the domain Ω. This problem has presented itself across numerous disciplines, including star-shaped domains in medical imaging [27] and synthetic aperture radar (SAR) [28], in which the demodulated SAR return signals from the moving antenna are approximately samples of 1-D Fourier transforms of projections of the scene reflectivity function. In this case, these projections are polar-grid samples of the 2-D FT of the scene. Figure 2.3(b) illustrates a typical case of a high-resolution imaging scheme that collects samples of the Fourier transform along radial lines at a relatively few number of angles (256 samples along each of 22 radial lines) for the Logan-Shepp phantom test image of Figure 2.3(a).

Extensive work both in the medical imaging and radar image processing communities have focused on reconstruction of an object from polar frequency samples using filtered back-projection algorithms [9]. The caveat of the algorithm is the assumption that all unobserved Fourier coefficients are zero, in essence reconstructing an image of minimal energy using the observation constraints. Figure 2.3(c) details the severe deficiencies of this approach, producing non-local artifacts that make basic image processing techniques difficult. The desire to accurately interpolate the missing values proves even more difficult due to the oscillatory nature of the Fourier transform.

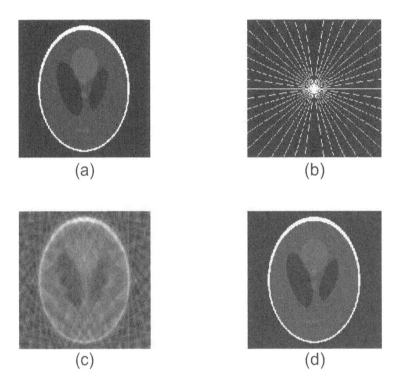

Figure 2.3: Example of image reconstruction using the Logan-Shepp phantom image: (a) original 256×256 image; (b) 22 radial samples in Fourier domain; (c) minimum-energy reconstruction; and (d) total variation reconstruction (nearly exact replica).

In their ground-breaking work on compressive sensing, Candes et al. [29], however, propose the use of convex optimization under the assumption that the gradient is sparse [29]. Letting $x_{i,j}$ denote a pixel in the ith row and j^{th} column of an $n \times n$ image X, the total variation of the image can be defined as the sum of the magnitudes of the discrete gradient at every point:

$$\text{TV}(x) = \sqrt{\left(x_{i+1,j} - x_{i,j}\right)^2 + \left(x_{i,j+1} - x_{i,j}\right)^2} . \tag{2.16}$$

Similar to the way we defined the sparse representation reconstruction in terms of a tractable l_1-minimization problem, the problem of image recovery can be recast as a second order cone problem (SOCP),

$$\min \text{TV}(x) \text{ subject to } y = Dx \tag{2.17}$$

where again D is the measurement matrix (a.k.a dictionary) and y is a small set of linear measurements. The SOCP convex optimization problem can be solved with great efficiency using interior point methods. Figure 2.3(d) shows the exact replica of the test image obtained using the convex optimization solvers contained in the l_1-Magic [25] package. The package additionally contains a demo script to compute the reconstructed image, of which we have modified to generate the images in Figure 2.3 and included in Appendix A.2.

CHAPTER 3

Dimensionality Reduction

Dimensionality reduction refers to methods in which high-dimensional data is transformed into more compact lower-dimensional representations. These techniques have been proven to be particularly useful in signal compression, discovering the underlying structure of the high-dimensional data and in providing the necessary understanding of patterns for data analysis and visualization. Ideally, the compact representation has a dimensionality that corresponds to the intrinsic dimensionality of the data, i.e., the dimensionality corresponding to the minimum number of free parameters of the data. In developing an in-depth and detailed understanding of the underlying structure of the data, methods to facilitate classification, visualization, and compression of high-dimensional data can be developed.

The problems encountered with dimensionality reduction have also been addressed using a technique known as manifold learning. When considering data points in a high-dimensional space, it is often expected that these data points will lie on or close to a low-dimensional non-linear manifold [30]. The discovery of the underlying structure of the manifold from the data points, assumed to be noisy samples from the manifold, is a very challenging unsupervised learning problem. The non-linear techniques developed attempt to preserve either the global or local properties of the manifold in the low-dimensional embedding. Consistent across all techniques however is the dependence on noise mitigation, needed to correct for or restrict data point outliers that are sufficiently outside the manifold.

The traditional linear dimensionality reduction methods, such as principal component analysis (PCA) [31, 32] and linear discriminant analysis (LDA) [33], have given way to non-linear techniques that have been shown to be more successful in discovering the underlying structure for highly non-linear data sets, e.g., the Swiss data roll. While the linear techniques were unsuccessful in determining the true embedding dimension for artificial data sets, they have been shown to outperform the non-linear techniques for real-world data sets [34]. The reasons for the poor performance of the non-linear algorithms include difficulties in model parameter selection, local minima, and manifold overfitting. Going forward, there seems to be increased interest in generating non-linear models that account for the geometry of the manifold using linear models [35].

In this chapter, a brief review of the popular dimensionality reduction/manifold learning methods will be reviewed. For the purposes of illustration, suppose we have an $n \times D$ matrix X consisting of n datavectors x_i each of dimensionality D. Assume that the intrinsic dimensionality of the data is known to be $d << D$. The dimensionality reduction techniques discussed will transform the dataset X into a reduced form Y of dimensionality $n \times d$, while retaining the geometry of the embedded manifold as much as possible. Neither the geometry of the original or embedded manifolds is known, nor is the intrinsic dimensionality of the datasets. Therefore, the problem of dimensionality reduction is typically an ill-posed problem that can only be solved by assuming some properties of the data [34].

The techniques for dimensionality reduction can roughly be classified into two groups: linear and non-linear. The linear techniques discussed in Section 3.1 assume the data lie on or near a linear subspace of the high-dimensional space. The non-linear techniques discussed in Section 3.2 do not make such an assumption and can thus form more complex embeddings of the high-dimensional data. Section 3.3 will present an emerging dimensionality reduction technique known as random projections that have become vital in compressive sensing. The use of the techniques is presented along with the discussion on sparse representations to show the correspondence between the two approaches. Examples of the lower-dimensional representations for radar imagery are presented in the final section as motivation for the use of sparse representations in radar image classification to be presented in Chapter 5.

3.1   LINEAR DIMENSIONALITY REDUCTION TECHNIQUES

The two well-known techniques for linear dimensionality reduction are principal component analysis (PCA) and linear discriminant analysis (LDA). As mentioned, both techniques assume a linear subspace of the high-dimensional data input space and as such, perform well when the underlying manifold is a linear or affine subspace of the input space. Note that a vector in the D-dimensional input space will be denoted by x_i and its low-dimensional counterpart by y_i, where the subscript i denotes the ith row of the input data matrix X.

3.1.1   PRINCIPAL COMPONENT ANALYSIS (PCA) AND MULTIDIMENSIONAL SCALING (MDS)

PCA is an extremely common and useful technique often used not only for dimensionality reduction but data visualization and feature extraction. Bishop [36] describes two different definitions for PCA: (1) maximization of the variance of the projected data using an orthogonal projection onto a lower-dimensional subspace (the principal subspace) and (2) minimization of the average projection cost, defined as the mean squared distance between the data points and their projections [37].

Mathematically, if we define the matrix S to be the covariance matrix for the data set X, PCA involves finding the d eigenvectors (i.e., principal components) of the covariance matrix. In other words, PCA solves the eigenproblem

$$SM = \lambda M \, , \tag{3.1}$$

where M is the sorted matrix of the d column eigenvectors corresponding to the d largest eigenvalues. The low-dimensional data representations are then computed by mapping them onto the linear basis, i.e., $Y = XM$.

The principal components can be determined in an incremental fashion to mitigate the large computational costs required for eigendecomposition of large covariance matrices. Efficient techniques such as the power method or EM algorithm can reduce PCA computation time by a factor of D (the dimensionality of the data set) [36]. Additionally, iterative techniques such as simple PCA [38] and probabilistic PCA [39] may be employed for approximation of the principal eigenvectors.

To see the usefulness of PCA, consider the data set and results shown in Figure 3.1 generated using script Program 3.1. It is obvious that a linear trend exists between the random variable x and the random variable y. Using the supplied script, the principal eigenvectors are computed using the MATLAB® function `princomp()` and have been overlaid onto the data set for easy analysis. Although not specifically shown, it can easily be verified using the 'latent' variable that the first principal component returned in the 'coeff' structure is the first principal eigenvector and is aligned with the linear trend of the data set. Similarly, the slope of this eigenvector is nearly identical to that used to generate the line. Intuitively, to maximize the variance between the projected points, we would like to see an eigenvector with the exact slope of our designed line so that the distance between the projected points is maximized.

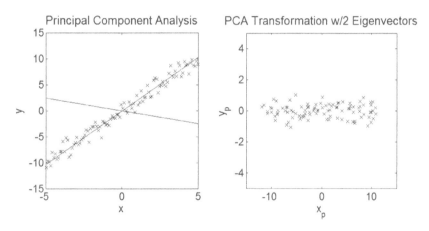

Figure 3.1: Principal component analysis for noisy points sampled from a line. Left: principal eigenvectors superimposed onto data set. Right: projection onto eigenvectors.

Similarly, multidimensional scaling (MDS) allows for efficient dimensionality reduction by attempting to maintain the original pairwise distances between points in the data set after reduction. Numerous stress functions exist for MDS which take into account the desire to maintain distances between differing sets of points. The raw stress function (3.2) for example weights the projected distance error the same for all data points. The Samson stress function however places more emphasis on maintaining the smaller distances between points in the input space. All stress functions, however, rely on simple and efficient techniques for minimization of the pairwise distance matrix, including popular techniques such as eigendecomposition, the conjugate gradient method, and the pseudo-Newton method [34]:

$$\phi(\boldsymbol{Y}) = \sum_{ij}(\|\boldsymbol{x}_i - \boldsymbol{x}_j\| - \|\boldsymbol{x}_i - \boldsymbol{y}_j\|)^2 \ . \tag{3.2}$$

```
clear all; close all; clc

%Global variables
var = 1; slope = 2;

%Generate data set
x = linspace(0,10,100)';
y = slope*x + sqrt(var)*randn(length(x),1);
X = [x y];

%Normalize data (subtract mean across each dimension)
Y = X - repmat(mean(X),size(X,1),1);

%Compute principal components
[coeff,score,latent] = princomp(X);

%Plot the results and overlay the principal components
figure;
plot(Y(:,1),Y(:,2),'kx');hold on;
xlabel('x');ylabel('y');title('Principal Component Analysis');

%Determine the two endpoints for each line
m = coeff(2,:)./coeff(1,:);           %Slope of line
xhat = mean(x);                       %Mean of x
yl =  m'*[xhat -xhat];
plot([xhat xhat; -xhat -xhat], yl','k-');

%Project onto single principal component
Yproj = coeff'*Y';
figure;
plot(Yproj(1,:),Yproj(2,:),'kx');axis([-15 15 -xhat xhat]);
xlabel('x_p');ylabel('y_p');title('PCA Transformation');
```

Program 3.1: PCA analysis.

3.1.2   LINEAR DISCRIMINANT ANALYSIS (LDA)

Originally developed in 1936 by R.A. Fisher, discriminant analysis was designed primarily as a classification tool but has since been used in dimensionality reduction. The aim of the algorithm was to find a linear mapping M that maximized the linear class separability in the low-dimensional representation of the data. In other words, the basic idea was to maximize a function that gives a large separation between projected class means while also giving a small variance within each class to consequently minimize class overlap [36].

As with all classification architectures, perfect class separability is typically not feasible. The goal of Fisher's algorithm was to optimize the ratio of the between-class variance to within-class variance. Defining the between-class variance as S_b and the within-class variance as S_w, the optimization is found using a linear mapping that maximizes the Fisher criterion:

$$\phi(M) = \frac{M^T S_b M}{M^T S_w M} \ .$$
(3.3)

For the purposes of dimensionality reduction, the low-dimensional data representation for the datapoints X are computed by mapping them onto the linear basis M. It is important to note, however, that unlike PCA, LDA is a supervised learning algorithm and requires the knowledge of the number of classes. For this reason, as a dimensionality reduction tool, LDA is not as common as its linear cousin, PCA. For MATLAB® implementations of the LDA algorithm, refer to the documentation for the 'classify' function.

3.2 NONLINEAR DIMENSIONALITY REDUCTION TECHNIQUES

The linear techniques discussed in the preceding section are established and well-understood techniques for dimensionality reduction. These techniques are able to discover the true structure of high-dimensional data if it lies on or near a linear subspace of the input space. Consider, however, a highly non-linear case, such as data sampled from the "Swiss Roll" shown in Figure 3.2 and generated using the algorithm in Program 3.2. Linear techniques effectively use Euclidian distance between points to determine embedding. Unfortunately, points lying far apart on the non-linear, lower-dimensional manifold may be much closer in the higher-dimensional input space, as measured by their straight-line Euclidian distance.

Non-linear techniques attempt to mitigate the issues encountered when using linear dimensionality reduction algorithms on non-linear manifolds. These techniques continue to see an increase in development as additional avenues for research are explored, including studies in human vision, speech, and motor control [7]. Technique variations are numerous but Maaten [34] classifies nonlinear dimensionality reduction techniques into three broad categories: (1) global preservation of original data; (2) local preservation of original data; and (3) global alignment of a number of linear models. Recent research [40] has also focused on methods that attempt to mitigate the effect of noise which can drastically change the manifold structure. Known as neighborhood smoothing embedding, the method can be used as a preprocessing technique to improve the performance of nonlinear dimensionality reduction techniques. For the purposes of brevity and demonstration, only a single technique from each category will be discussed while references for similar techniques will be given.

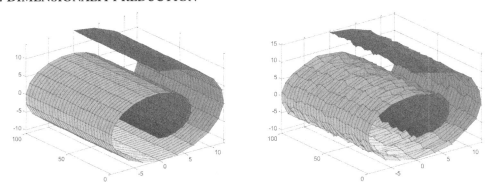

Figure 3.2: "Swiss roll" manifold with noisy version on right (N = 1000, ht = 100, var = 2).

```
function data = swissRoll(N, ht, var)

tt0 = (3*pi/2)*(1+2*linspace(0,1,sqrt(N)));
hh = linspace(0,1,sqrt(N))*ht;
xx = (tt0.*cos(tt0))'*ones(size(hh));
yy = ones(size(tt0))'*hh;
zz = sqrt(var)*rand(size(yy))+((tt0.*sin(tt0))'*ones(size(hh)));
cc = tt0'*ones(size(hh));
surf(xx,yy,zz,cc);
axis tight;

data = [xx(:) yy(:) zz(:) cc(:)];
```

Program 3.2: "Swiss roll" data generation.

3.2.1   ISOMAP

The ability of the human nervous system to extract relevant features from over 30,000 auditory or 10^6 optic nerve fibers is an excellent demonstration in dimensionality reduction. Consider pictures of a human face observed under different pose and lighting conditions. The high-dimensional images lie on an intrinsically three-dimensional manifold that can be parameterized using just two pose variations and a single lighting angle [7].

While the linear techniques are guaranteed to recover the true structure of the data given that it lies on or near a linear subspace, ISOMAP provides improvements by allowing for the flexibility to learn the intrinsic geometry of nonlinear manifolds. Additionally, ISOMAP retains the advantageous algorithmic features of the linear techniques, such as implementation ability, computational tractability, and the convergence guarantees of the linear learning methods.

The basic tenet of the algorithm is the use of geodesic distances, or shortest path distances along the manifold, rather than Euclidian distances. As mentioned with the Swiss roll example, these can be markedly different from Euclidian distances for non-linear manifolds. The difficulty in the algorithm lies in estimating the geodesic distance given only input-space distances. For neighboring points, the geodesic distance can be estimated using input-space distance. For far away points, this can be approximated by adding up a sequence of "short hops" between neighboring points [7]. Spectral graph theory allows for the efficient computation of these distances by computing the shortest paths in a graph with edges connecting neighboring data points.

The weighted graph G is computed by identifying neighboring points (those within a fixed radius ϵ or its K nearest neighbors) based on input-space distance. Edge weights between neighboring points are determined by the input-space distance. The geodesic distances are then estimated by computing the shortest path distances in the graph G from one point to another. Classical MDS can be used to construct an embedding of the data in d-dimensional Euclidian space.

ISOMAP is a non-iterative, polynomial time procedure that retains the global characteristics of the manifold, i.e., the geodesic distances between points on the manifold. As with PCA or MDS, the algorithm allows for the discovery of the true dimensionality of the data set by estimating the "elbow" in the error decrease as the embedding dimension is increased. Additionally, in the limit of infinite data, the geodesic distance approximations become arbitrarily accurate, guaranteeing asymptotic convergence.

Similar to ISOMAP, maximum variance unfolding (MVU) defines a neighborhood graph on the data and retains the pairwise distances [41]. By maximizing the Euclidean distance between the datapoints in the embedding, while retaining the distances in the neighborhood graph, MVU can efficiently "unfold" the manifold, again without altering the local geometry, using basic semidefinite programming techniques. Other methods that retain the local manifold structure include diffusion maps [42], and kernel PCA [43], which is an extension of the original linear PCA algorithm using kernel functions. Reformulation of PCA in kernel space (essentially the inner-product of the datapoints in high-dimensional space) allows for the construction of nonlinear mappings.

3.2.2   LOCAL LINEAR EMBEDDING (LLE)

Local linear embedding is an unsupervised learning algorithm that preserves neighborhood relationships for high-dimensional inputs. Unlike clustering methods however, LLE maps its inputs into a single global coordinate system of lower dimensionality and its optimizations do not involve local minima [44]. The algorithm is seen as an improvement over techniques like multidimensional scaling (MDS) [45] and the previously mentioned ISOMAP that attempt to preserve the pairwise distances or geodesic distances between all points in the data set. In doing so, LLE becomes less sensitive to short-circuiting caused by the estimation of Euclidian or geodesic distances between widely separated points.

The basic assumption of the LLE algorithm is that the data, consisting of N real-valued vectors $\boldsymbol{x}_i$, $i = 1, \ldots, N$, each of dimensionality D, are sampled from some underlying manifold.

Each data point and its neighbors are assumed to lie on or close to a locally linear patch of the manifold. The local properties of the manifold can then be estimated by treating each data point as a linear combination of its nearest neighbors, essentially fitting a hyperplane to each data point and its neighbors [34].

The reconstruction weights for each data point are computed using the K nearest neighbors for that data point while requiring that the sum of the weights (rows of the weight matrix W) sum to one, i.e., $\sum_j w_{ij} = 1$. When formulated as such, the solution for the reconstruction weights is found by solving a least squares problem [44]. Moreover, since the reconstruction weights reflect the intrinsic properties of that data and are invariant to translation, rotation, and scaling transformations, the reconstruction weights for each data point in D dimensions also reconstructs its embedded manifold coordinates in d dimensions. Each high-dimensional observation $\mathbf{x}_i$ is mapped to a low-dimensional vector $\mathbf{y}_i$ by choosing the d-dimensional coordinates $\hat{\mathbf{y}}_i$ that minimize the embedding cost function

$$\Phi(\mathbf{Y}) = \sum_i |\hat{\mathbf{y}}_i - \sum_j w_{ij} \hat{\mathbf{y}}_j|^2 . \tag{3.4}$$

The coordinates of the low-dimensional representations $\mathbf{y}_i$ that minimize this cost function can be found by computing the eigenvectors corresponding to the smallest d non-zero eigenvalues of the matrix $\mathbf{I} - \mathbf{W}$, where $\mathbf{I}$ is the $N \times N$ identity matrix.

Sample results generated using script Program 3.4 and the function provided in script Program 3.3 for the Swiss roll dataset are shown in Figure 3.3. Using an embedding dimension of 2 and 12 nearest neighbors, the intrinsic 2-D manifold for the 2000 point data set can be fairly well estimated. As with ISOMAP, the true embedding dimension can be estimated by locating the dimensionality d in which no appreciable variance increase is noticed.

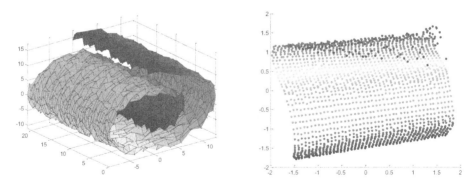

Figure 3.3: 2-D embedding of Swiss roll dataset using LLE.

Additional techniques that maintain local properties of the data include Laplacian eigenmaps [46], wherein cost function weights are inversely proportional to the distance between neighbors in the input space, Hessian LLE [47] which minimizes the "curviness" of the high-dimensional manifold in the low-dimensional embedding and local tangent space alignment (LTSA) which at-

```
function Y = lle(X,K,d)

[D,N] = size(X);

% Compute pairwise distance
X2 = sum(X.^2,1);
dist= repmat(X2,N,1)+repmat(X2',1,N)-2*(X'*X);

% Compute neighbors
[sorted,index] = sort(dist);            %Sort along columns
nhbd = index(2:(1+K),:);                %Ignore index to itself

% Compute reconstruction weights
if(K>D)       %Regularization will be used for ill fits
  tol=1e-3;
else
  tol=0;
end

W = zeros(K,N);
for ii=1:N
   z = X(:,nhbd(:,ii))-repmat(X(:,ii),1,K);          % shift ith pt to origin

   C = z'*z;                                         % local covariance

   C = C + eye(K,K)*tol*trace(C);                    % regularlization (K>D)

   W(:,ii) = C\ones(K,1);                            % solve Cw=1
   W(:,ii) = W(:,ii)/sum(W(:,ii));                   % enforce sum(w)=1

end;


% Compute embedding
M = sparse(1:N,1:N,ones(1,N),N,N,4*K*N);
for ii=1:N
   w = W(:,ii);
   jj = nhbd(:,ii);
   M(ii,jj) = M(ii,jj) - w';
   M(jj,ii) = M(jj,ii) - w;                          % Make symmetric
   M(jj,jj) = M(jj,jj) + w*w';
end;

% Calculate embedding
options.disp = 0; options.isreal = 1; options.issym = 1;
[Y,eigenvals] = eigs(M,d+1,0,options);
Y = Y(:,2:d+1)'*sqrt(N);

data = [xx(:) yy(:) zz(:) cc(:)];
```

Program 3.3: LLE algorithm.

tempts to align the linear mappings from the low- and high-dimensional spaces to a local tangent space [48].

```
LLE generation code
X = swissRoll(2000,21,10);
Y = lle(X(:,1:3)',12,2);
figure;scatter(Y(1,:),Y(2,:),12,X(:,4),'filled')
```

Program 3.4: LLE for Swiss roll dataset code.

3.2.3   LINEAR MODEL ALIGNMENT

The final approach to non-linear dimensionality reduction involves the use of global alignment for a number of linear models. These methods have been designed to address the shortcomings of global and local data preservation techniques. These methods include local linear coordination (LLC) [49] and manifold charting [50] among others. These approaches have been successfully applied in manifold analysis for facial recognition and handwritten digits.

LLC is employed by first computing a mixture of factor analyzers or probabilistic PCA components using the expectation-maximization (EM) algorithm. The local linear models are used to develop independent representations to which each datapoint has an associated responsibility. It is shown in [49], that using the same approach adopted in LLE, incorporating the identity matrix and the defined weight matrix W, that the alignment of the linear models can be performed by solving a generalized eigenproblem. The eigenvectors found are a linear mapping from the responsibility weighted representation to the low-dimensional data representation.

3.3   RANDOM PROJECTIONS

High-dimensional data sets continue to emerge with new applications and with computer capabilities on the increase. These data sets, namely text and images, can be represented as points in a high-dimensional space. The dimensionality often imposes limitations on conventional data processing methods such as those discussed in the preceding sections. The statistically optimal (at least in a mean-squared error sense) reduction is accomplished using PCA. Capturing as much of the variation in the data as possible comes at an increasingly high computational cost; the cost of determining the projection onto the lower-dimensional orthogonal subspace scales as the cubic of the dimension of the data set [51].

Random projection (RP) operators project high-dimensional data onto a lower-dimensional subspace using a purely random matrix. The Johnson-Lindenstrauss lemma provides the basis for the use of random projections for dimensionality reduction: a set of n points in a high-dimensional Euclidian space can be mapped down into an $\mathcal{O}(\log \frac{n}{\epsilon^2})$ dimensional Euclidian space such that the distance between any two points changes by only a factor of $(1 \pm \epsilon)$ [52]. The importance of this

```
clear; clc; close all;
figure;

%Number of DCT coefficients to use
k = 2000;

im =
imageRead('C:\DATA\MSTAR\TARGETS\TRAIN\17_DEG\T72\SN_132\HB03814.015'
);

subplot(1,3,1);imagesc(db(im));colormap(gray);axis('image');axis off;
im = imresize(im,[64 64]);
[m n] = size(im);
X = dct2(im);

[vals inds] = sort(abs(X(:)),'descend');
Xhat = zeros(size(X));

Xhat(inds(1:k)) = X(inds(1:k));
imhat = idct2(Xhat);

subplot(1,3,2);imagesc(db(imhat));colormap(gray);axis('image');axis
off;

%Now use random projections
rp = randn(k,m*n); rp = rp./(repmat(sum(rp),k,1));
Xrp = rp*double(im(:));
irphat = rp'*Xrp; irphat = reshape(irphat,[m n]);
subplot(1,3,3);imagesc(db(irphat));colormap(gray);axis('image');axis
off;
```

Program 3.5: Image compression and reconstruction using DCT and RP.

result is that interpoint distances are preserved when projected onto a randomly selected subspace of suitable dimension.

Mathematically, random projection of the original d-dimensional data $X \in \mathbb{R}^{d \times N}$ to a k-dimensional ($k << d$) subspace through the origin, using a random $k \times d$ matrix R is given by

$$\hat{X} = RX \, . \tag{3.5}$$

It should be noted that (3.5) is a linear mapping and not necessarily a projection as in general R is not orthogonal and can introduce significant distortions. Orthogonalizing R is very expensive but as noted in [53], for a high-dimensional space, there exists a much larger number of almost orthogonal than orthogonal directions. Thus, vectors generated randomly in high-dimensional spaces can be considered approximately orthogonal and the mapping to be a projection.

Bingham and Mannila [53] present distance distortion performances for text and image data using RP and other popular dimensionality reduction techniques. In addition to being significantly faster than its conventional counterparts, RP proved to not distort data significantly more than PCA. More importantly, RP proved adept at providing greater accuracies for significantly lower dimensions.

It should be noted that using random projections as a dimensionality reduction technique is primarily, although not exclusively, beneficial in applications where the distances of the original high dimensional data points are meaningful. As an illustration, consider the alternative use for random projections as an image compression technique, akin to the discrete cosine transform (DCT). Bingham [53] showed the superior performance of RP over the DCT for dimensionality reduction of text and data but Figure 3.4 clearly shows that the DCT is able to capture and reconstruct at least some vital information necessary for interpretation of a radar image by the human eye. (*Note:*

Figure 3.4: Left: Original MSTAR [54] SAR dB domain image. Middle: image reconstruction after using largest 2500 DCT components. Right: image reconstructions after random projection to a 2500-dimensional subspace.

The pseudoinverse of the random projection matrix R needed for reconstruction is expensive to compute, but since R is almost orthogonal, the transpose of R is a sufficient approximation.) Random projections discard vital signal information at the expense of maintaining interpoint distances in the lower dimensional subspaces. Distance preservation however becomes important for tasks such as training neural networks using clustering or k nearest neighbors or techniques that rely heavily on interpoint distances. As such, the use of random projections for dimensionality reduction should be approached with care.

CHAPTER 4

Radar Signal Processing Fundamentals

By radiating energy into space and analyzing the echo reflection signals, radar systems have the ability to detect all types of objects as well as determine their distance and speed relative to the radar system. The ability to do so is rooted in basic signal processing fundamentals.

Consider the block diagram for a pulsed monostatic radar shown in Figure 4.1. The electromagnetic signal transmitted by the antenna is reradiated by the target(s) in the scene back to the radar. The received signal is then processed by both hardware and software modules to extract the information that is useful to the system. For early radar systems, this included manual extraction of target presence along with information about its range, angular location and relative velocity. The range, or distance, to a target is proportional to the total propagation time for the transmitted and reflected signal to return to the receiver. Angular location can be computed by considering the angular direction of the receiver corresponding to the maximal amplitude of the return signal at the instant a target is detected. Doppler frequency shifts in the echo signal are a result of the relative motion between the radar platform and the target, indicating the targets radial velocity. These three fundamental measurements have allowed for application extensions that early radar engineers may never have imagined.

This chapter will provide the very basics in radar signal processing needed to understand the emerging radar technologies that have incorporated both sparse representations and compressive sensing. For advanced details of radar signal processing techniques, consider the texts by both Skolnik [4] and Richards [2].

4.1 ELEMENTS OF A PULSED RADAR

A sample block diagram for a conventional monostatic pulsed radar is shown in Figure 4.2. While this layout is by no means unique, it does allow for the identification of areas within the system in which sparse representations have the potential to be included. The location for the digitization of the analog signal in current digital radar systems has been purposefully left out. Early digital systems had embedded the A/D converter within the signal processing unit, but advancements have allowed for digitization of the signal at the IF stage, moving the A/D converter closer to the radar front end. Sparse representations have seen implementations both before and after the A/D conversion. The highlighted system areas show this capability to operate either directly on an analog IF signal (signal

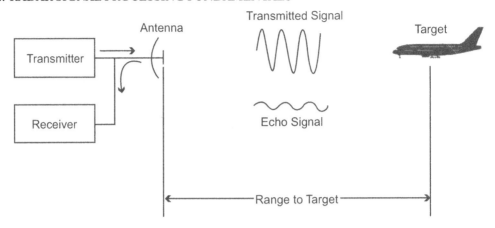

Figure 4.1: Basic principle of radar. (Based on Jakowatz, et al., *Spotlight-Mode Synthetic Aperture Radar: A Signal Processing Approach*. New York: Springer Science + Business Media, 1996. Copyright © 1996, Springer Science + Business Media [28].)

processing) or on the converted and perhaps already extracted radar data itself (data processing). Examples of each technique will be discussed in Chapter 5.

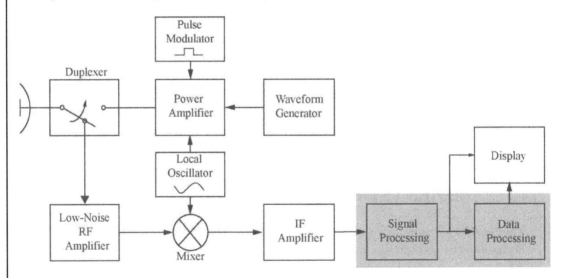

Figure 4.2: Block diagram for conventional pulsed monostatic radar.

There are essentially three key areas in a pulsed radar system. The transmitter and waveform generator properties, such as transmission frequency, are vital in the tuning the sensitivity and range resolution of the radar. Most radars operate in the microwave frequency region from 200 MHz to about 95 GHz. Transmission frequencies are selected based on the system requirements for transmission power, atmospheric attenuation and antenna size. The transmitted signal is typically a *pulse-train* (hence the name pulsed radar) and contains a series of shaped pulses modulating a sinewave carrier, as shown in Figure 4.3. The pulse width $\tau_b = 1\,\mu s$, pulse repetition period $T_p = 1$ ms and the peak power is $P_t = 1$ MW represent values comparable to those seen in a medium-range air-surveillance radar [4]. These systems can operate at peak powers up to 10 MW with average powers below 10–20 kW. As with any system, there are important design trade-offs that must be considered. Detection performance increases with transmission power but range resolution decreases as the pulse length increases. This is discussed later in this chapter.

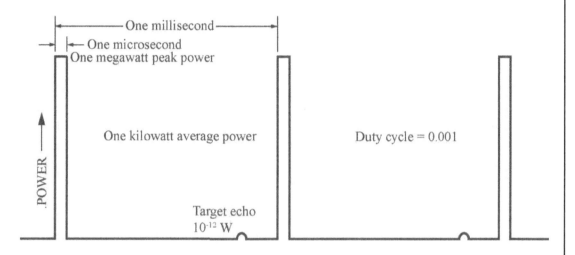

Figure 4.3: Typical pulse waveform values for a medium-range air-surveillance radar. The rectangular pulses represent pulse-modulated sinewaves.

The second element of a pulsed radar system is the antenna itself. Angular resolution is a function of the main lobe width for the signal transmitted by the antenna. Smaller beamwidths require larger apertures or shorter transmission wavelengths. In addition, side lobe levels must be kept at a minimum to reduce the effect echoes from nearby scatterers have on one another. Synthetic aperture radar, presented in Section 4.3, overcomes these obstacles by synthesizing a larger antenna aperture to improve angular resolution for the formation of a 2-D image. The third and final element, the radar receiver located in the bottom portion of Figure 4.2, is responsible for the demodulation of the return signal to baseband and information extraction for further manual or automated processing. Typically, the received signal is split into two channels: the *in-phase* and *quadrature* phase channels so as to remove any ambiguity in the received signals phase. Once the signal has been successfully demodulated, signal processing algorithms like moving target indicators (MTIs) can be implemented to locate moving targets. Similarly, data processing algorithms like automatic target recognition systems (ATRs) attempt to identify a target using a 2-D radar image.

4.2 RANGE AND ANGULAR RESOLUTION

Resolution refers to the ability of the radar to distinguish between targets. *Range resolution* specifically is the ability to identify two or more targets on the same bearing but at different ranges (straight-line distance from the radar). Similarly, *angular resolution* concerns target identification in both the azimuth (2nd) and elevation (3rd) dimensions. Resolution capabilities are controlled by the transmitted waveform properties and antenna beamwidths, respectively. A simple, intuitive introduction to the two concepts follows in this section. For more complete discussions, see Richards [2], Skolnik [4], or Jakowatz [28].

To analyze range resolution, consider a standard pulse waveform element for the train in Figure 4.3:

$$s(t) = b(t) \cos(\omega_0 t) . \tag{4.1}$$

The envelope function $b(t)$ (usually rectangular, hamming, raised cosine, etc.), with a duration of τ_b seconds, modulates a carrier wave at a frequency ω_0. An example waveform generated using the raised cosine envelope function $b(t) = .5[1 + \cos(\frac{2\pi t}{\tau_b})]$ is shown in Figure 4.5(a). The time t_R it takes to receive the return signal for a target at range R is

$$t_R = \frac{2R}{c} , \tag{4.2}$$

where $c \approx 3 \times 10^8$ m/s is the speed of light or the rate electromagnetic energy travels through free space. Assuming a constant-frequency pulse is transmitted at time $t = 0$ for τ_b seconds, for two targets at ranges R_1 and R_2, the leading edge of the return pulse for each signal will be received at times t_1 and t_2 respectively. For the constant-frequency pulse, the system will only be able to discern between the two echo signals if there is no overlap in the return echoes. Thus, the range resolution $\Delta R = R_2 - R_1$ is determined by finding the minimum distance the targets must be separated to prevent signal return overlap:

$$t_2 - t_1 = \tau_b$$

$$\frac{2R_2}{c} - \frac{2R_1}{c} = \tau_b \tag{4.3}$$

$$\frac{2}{c}(R_2 - R_1) = \frac{2}{c}\Delta R = \tau_b$$

$$\Delta R = \frac{\tau_b c}{2} .$$

This point is illustrated in Figure 4.4. We can similarly discuss the range resolution from the signal bandwidth perspective. For rectangular and shaped pulse waveforms the time-bandwidth product is always equal to unity in cycle measure ($\tau_b B = 1$) [28]. Recall that the bandwidth of the transmitted pulse is inversely proportional to the pulse length. This implies that as the pulse length increases, the bandwidth of the transmitted signal decreases as shown in Figure 4.5(c). Substituting $B = 1/\tau_b$ into (4.3), we see that range resolution decreases as the bandwidth increases, i.e.,

$$\Delta R = \frac{c}{2B} \, . \tag{4.4}$$

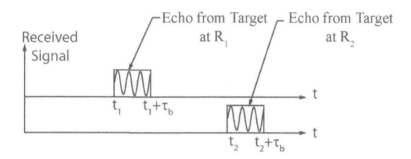

Figure 4.4: Geometry for describing range resolution. For constant frequency pulses, there must be no overlap between the two return signals. (Based on Richards, *Fundamentals of Radar Signal Processing.* New York: McGraw-Hill, 2005. Copyright © 2005, McGraw-Hill, [2].)

Improved range resolution thus requires either a shorter pulse (reducing the average transmit power) or increased bandwidth. Luckily, there are alternative ways to encode radar signals to increase bandwidth to improve range resolution while not suffering from the low average power levels that plague short continuous-wave (CW) bursts. The most common of these *stretched* or *dispersed* [28] waveforms is the linear FM chirp

$$s(t) = \text{Re}\{\exp[j(\omega_0 t + \alpha t^2)]\} \, . \tag{4.5}$$

The FM chirp contains a linearly increasing frequency component based on the so-called *chirp rate* of 2α. As shown in Figure 4.5(b), the frequencies encoded by the chirp extend from $\omega_0 - \alpha \tau_b$ to $\omega_0 + \alpha \tau_b$, resulting in a much larger effective bandwidth of approximately

$$B_c = \frac{\alpha \tau_b}{\pi} \, . \tag{4.6}$$

The increase in bandwidth is obvious from the chirp frequency spectrum shown in red in Figures 4.5(c) and 4.5(d). The FM chirp has the capability to transmit higher average power signals using longer pulse lengths while maintaining the large bandwidths needed for high range resolution. Known as *pulse compression*, the FM chirp produces range resolution properties of a pulse with a duration that is equivalent to the inverse of its bandwidth [28].

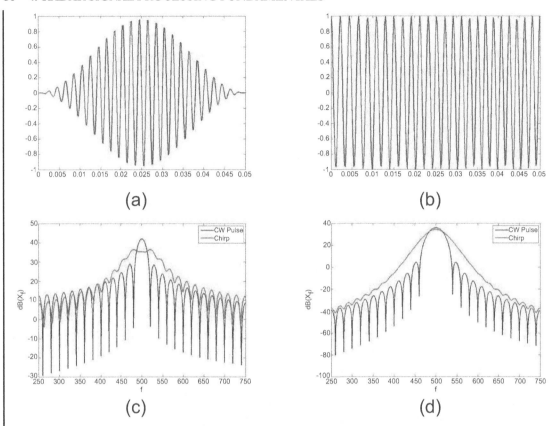

Figure 4.5: (a) Continuous wave pulse raised cosine envelope function. (b) Linear FM chirp waveform. (c) Frequency spectrum for CW and linear FM signal using rectangular pulse. (d) Frequency spectrum for CW and linear FM signal using raised cosine envelope. The script used to generate all figures is contained in Appendix A.3.

As mentioned, angular resolution is the ability to distinguish between two targets at the same range but at different azimuth or elevation angles. Scatterer echoes will be combined at the receiver if the targets simultaneously lie in the main lobe of the illumination beam of the antenna. While the discussion on antenna properties is outside the scope of this text, it is worth noting that the estimation of the angular resolution is determined by locating the 3-dB beamwidth θ_3 of the antenna. The *cross-range resolution*, or the resolution in the dimension orthogonal to the range, can be estimated as

$$\Delta CR = 2R \sin\left(\frac{\theta_3}{2}\right) \approx R\theta_3 , \tag{4.7}$$

where the approximation holds when the 3-dB beamwidth is small, typical for pencil beam antennas [2]. This approximation is shown in Figure 4.6.

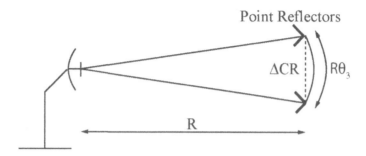

Figure 4.6: Angular resolution as a function of the 3-dB beamwidth of the antenna.

4.3   IMAGING

The idea of radar imaging brings to mind the 2-dimensional target indication screens known as plan-position indicators shown previously in Chapter 1. Radar systems also have the capability to produce high-resolution imagery, such as those created using a synthetic aperture radar (SAR) imaging system. A SAR image of Washington, D.C. is shown in Figure 4.7. The image itself is very similar to that of an optical image however some immediate differences can be noted. The monochromatic nature of the image makes object differentiation and detail extraction difficult. The image noise, or speckle, appears to significantly reduce the resolution of the photograph. These difficulties may seem to hinder the use of radar imagery but with current resolutions that approach that of optical imagery, radar images have the unique advantage of not only being able to be formed at long distances, but also in inclement weather and under the cover of night.

Of the radar imaging technologies available, of particular importance in this book is SAR imagery. The notion of range resolution was previously discussed and can be improved using various signal processing techniques. The cross-range resolution, however, for a *real aperture radar* is determined by the width of the antenna beam. For even nominal stand-off distances, conventional radar imaging systems would produce images with cross-range resolutions on the order of 100 meters, far too coarse for real-world use. Synthetic aperture radars synthesize the larger antenna aperture sizes needed to reduce the beamwidth by having the antenna move in relation to the area being imaged. These airborne or space-based radars transmit pulses along a desired path and after proper processing, produce extremely detailed imagery.

Consider the real-aperture radar system shown in Figure 4.8(a). For a moving imaging system transmitting and processing a single pulse, the width of the cross-range beam on the ground, and

Figure 4.7: 1-m resolution SAR image of Washington, D.C. (Image courtesy of Sandia National Labs, http://www.sandia.gov/radar/imageryku.html.)

subsequently the cross-range resolution, is given by

$$\Delta CR = \beta R = \frac{R\lambda}{D} \, , \tag{4.8}$$

where λ is the transmitted signal wavelength, D is the diameter of the physical aperture, and β is the angular beamwidth. The cross-range resolution is thus proportional to the range and inversely proportional to the size of the aperture. For an X-band radar operating at 10 GHz from a stand-off range of 10 km using a 1-m antenna, the cross-range resolution is only 300 m. Since decreasing the stand-off range is undesirable for many military scenarios and wavelength selection is a function of the desired electromagnetic properties for the transmitted signal, the only feasible way to increase cross-range resolution is to increase the aperture size. Since imaging antennas are often mounted on aircraft, the physical size is often limited to that of the imaging platforms payload capabilities.

The goal of the radar imaging system is to produce an estimate of the two-dimensional reflectivity function, $|g(x, y)|$. If we consider transmitting multiple pulses, as shown in Figure 4.8(b), all targets lying along the same constant-range contour will be received at the same instant. The received signal thus cannot be related to any particular cross-range/range position (x, y). Instead, the received signal is the integration of the reflectivity values from all targets that lie along an approximately straight constant range line. Given that the transmitted signal is a linear FM chirp, the received signal (after deramp processing) is of the form

$$r_c(t) = C \int_{-L}^{L} \left[\int_{-L}^{L} g(x, y) dx \right] e^{-j\phi} dy \, , \tag{4.9}$$

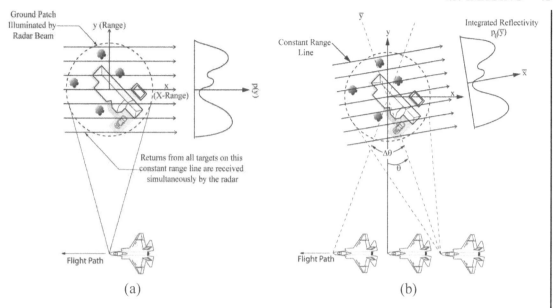

Figure 4.8: (a) Real aperture radar system. Angular beamwidth is determined by the ratio of the wavelength to the diameter of the physical aperture. (b) Synthetic aperture radar imaging concept. Multiple pulses along the flight path are processed to improve cross-range resolution. (Based on Jakowatz, et al., *Spotlight-Mode Synthetic Aperture Radar: A Signal Processing Approach.* New York: Springer Science + Business Media, 1996. Copyright © 1996, Springer Science + Business Media [28].)

where the inner integral indicates the estimate of the line integrals of $g(x, y)$ taken along the cross-range direction. The radar system processes many echoes over some interval of viewing angles, $\Delta\theta$. The corresponding deramped return at each interval is of the form

$$ r_c(t) = C \int_{-L}^{L} \left[\int_{-L}^{L} g(\bar{x}\cos\theta - \bar{y}\sin\theta, \bar{x}\sin\theta + \bar{y}\cos\theta)d\bar{x} \right] e^{-j\phi}dy \ . \qquad (4.10) $$

With the antenna ensured to point at the same ground patch center, at each viewing angle θ, information about the scene reflectivity is found along different line integrals. This range averaging is a projection, at least in the tomographic sense, of the two-dimensional scene into a one-dimensional function [2]. For the purposes of image construction, according to the *projection slice theorem* [55], the one-dimensional Fourier transform of the projection function $p_\theta(\bar{y}) = \int_{-L}^{L} g(\bar{x}\cos\theta - \bar{y}\sin\theta, \bar{x}\sin\theta + \bar{y}\cos\theta)d\bar{x}$ given by (4.10) is equal to the two-dimensional Fourier transform of the image to be reconstructed evaluated along a line in the Fourier plane that lies at the same angle θ. In other words, each slice provides a measurement of the two-dimensional Fourier transform of the scene along the same angle. These slices can then be interpolated to build up a complete Fourier transform of the scene reflectivity function.

Additional processing is required to produce a useful image including the use of polar formatting to transform the data from a polar to a rectangular grid. As it turns out, the degree of cross-range resolution in the reconstructions depends only on the transmission wavelength λ and on the size of $\Delta\theta$. The ability to sweep out measurements to synthesize a larger aperture and improve cross-range resolution has afforded SAR systems amazing capabilities that continue to see use in applications similar to those for their optical counterparts.

4.4   DETECTION

Target detection was one of the primary goals when engineers considered the first radar system. Detection refers to the process by which a decision is made as to whether the received signal is the result of an echo from a target or simply represents the effects of interference. The complexity of radar signals requires the use of statistical models due to the multitude of interference signals. Comprehensive models have been developed both for interference and echoes from complex targets. The processing of detection decisions then is a problem in statistical hypothesis testing. When testing for the presence of a target, one of two hypotheses can be assumed to be true: the *null hypothesis* H_0 which assumes the signal is the result of interference only and the second hypothesis H_1 where the signal consists of both interference and target echo returns. Once the two conditional probability density functions

$$p_y(y|H_0) = \text{pdf of } y \text{ given that a target was } not \text{ present}$$
$$p_y(y|H_1) = \text{pdf of } y \text{ given that a target } was \text{ present}$$

have been defined, metrics such as the probability of detection, P_D, probability of false alarm, P_{FA}, and the probability of miss, $P_M = 1 - P_D$ can be defined. The *Neyman-Pearson* criterion can then be used to maximize the probability of detection under the constraint that the probability of false alarm does not exceed a set value [2]. Using this technique, the method of *threshold detection*, where a signal's amplitude above a set threshold indicates target presence, can be used for detection.

Figure 4.9 illustrates threshold detection for a 1-D range signal return. This threshold is determined based on a desired operating point, set for a specific false alarm or detection rate. There are of course numerous details in implementing a threshold detector not discussed. Various designs utilize the magnitude, squared-magnitude, or even log-magnitude of the complex signal samples. Constant false alarm rate (CFAR) detectors serve to provide an estimate for the threshold when interference statistics are not known. For a detailed discussion on detection theory with implementation examples, see Kay [56].

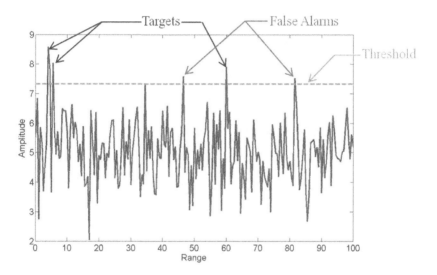

Figure 4.9: Threshold detection for range signal trace.

CHAPTER 5

Sparse Representations in Radar

While military applications such as surveillance, navigation, and weapons guidance dominate the current radar landscape and drive radar technology development, increasing civilian radar applications continue to rise. In addition to the technologies previously mentioned, radar has found use in pedestrian and vehicular collision avoidance systems, environmental mapping and even in the study of the movement of insects and birds. All these technologies rely on a continually evolving signal processing methodology. As such, the potential for radar to impact new civilian and military applications relies heavily on the use of innovative and efficient signal processing algorithms. Advanced algorithm architectures and new or improved radar technologies often go hand in hand. The emergence of sparse representation theory in the field of radar systems is no different. This chapter will serve as a direct example of the utility of the sparse representation framework throughout the field of radar.

Each section of this chapter will focus on a recently developed radar application that uses sparse representations and elements of compressive sensing in its signal processing architecture. The ability of sparse representations to find accurate and concise signal models allows for improvements in signal representation, compression and inference, as evidenced by this growing list of radar technologies. This of course is not an exhaustive list, as the number of available radar applications and implementations of the sparse representation framework seems to grow daily. The varied application techniques for sparse representations throughout the radar community should serve to display the power of the approach and potentially spawn ideas for future radar applications.

5.1   ECHO SIGNAL DETECTION AND IMAGE FORMATION

As mentioned in Section 4.2, to simultaneously achieve the energy of a long pulse and the resolution of a short pulse, pulse compression technology takes advantage of the fact that a long pulse can have the same spectral bandwidth as a short pulse if modulated in frequency or phase [4]. Unfortunately, the strain to process the ultra-wideband signal falls on the Analog-to-Digital Converters (ADCs), whose hardware capabilities remain the largest limiting factor in radar system developments [57]. Both existing and advanced radar systems can benefit from the ability to lower the required sampling rate. The reduced requirements for the ADC have the potential to not only improve hardware reliability but reduce development cost.

Consider the standard linear filter model for a received radar signal [4]:

$$y(t) = \int_{-\infty}^{\infty} x(t)s(t-\tau)d\tau + \eta(t) , \tag{5.1}$$

where $x(t)$ is the transmitted signal, $s(t)$ is the radar reflectivity response for the target, and $\eta(t)$ is noise. The received signal $y(t)$ is the convolution of the transmitted signal with the target impulse response. When discretized using N samples this becomes [58]

$$y[n] = \sum_{k=1}^{N} x[k]s[n-k] + \eta[n] = Xs + \eta , \qquad (5.2)$$

where $s, \eta \in \mathbb{R}^{N \times 1}$. The transmission signal matrix $X \in \mathbb{R}^{N \times N}$ contains time-shifted versions of the transmitted signal in each column. It is important to note that the matrix X is Toeplitz since each descending diagonal from left to right is constant.

The objective of the radar system is to accurately recover s from y. A simple solution is to estimate the signal that produces the least-squared error, i.e., compute s by minimizing the l_2-norm of the estimation error. Recall, this is found by computing the pseudo-inverse of the transmission signal matrix X. If it is known that the target scene is sparse, i.e., $\text{supp}(s) = S << N$, the least squares solution will be uninformative as energy will be distributed among all coefficients in s. Recall that the exact sparse solution can be recovered using convex optimization techniques, however, the large sampling rates result in extremely large data vectors and matrices, making solution rates lengthy if not impractical.

Compressive sensing [59] provides the capability of measuring y using $M < N$ measurements by projecting the signal onto a second set of basic vectors $\{\boldsymbol{\psi}_m\}, m = 1, \ldots, M$ by

$$z(m) = \langle y, \boldsymbol{\psi}_m^T \rangle , \qquad (5.3)$$

where $\boldsymbol{\psi}_m^T$ denotes the transpose of $\boldsymbol{\psi}_m$ and $\langle \cdot, \cdot \rangle$ refers to the inner product. In matrix notation, this is equivalent to

$$z = \boldsymbol{\Psi} y = \boldsymbol{\Psi} Xs + \boldsymbol{\Psi} \eta , \qquad (5.4)$$

where $\boldsymbol{\Psi} \in \mathbb{R}^{M \times N}$ is the measurement matrix with each row corresponding to a basis vector $\boldsymbol{\Psi}_m$ and z is an $M \times 1$ column vector. Compressive sensing relies on the near orthonormality of matrices satisfying the *restricted isometry property* when operating on sparse vectors. Mathematically, matrix $A = \boldsymbol{\Psi} X \in \mathbb{R}^{M \times N}$ is said to satisfy the restricted isometry property if for an integer $s < p$ and every submatrix A_s, there exists a constant δ_s for every vector u such that

$$(1 - \delta_s)\|u\|_2 \le \|A_s u\|_2 \le (1 + \delta_s)\|u\|_2 . \qquad (5.5)$$

This basic principle states the embedding of the p-dimensional vector in a random m-dimensional space does not severely distort the norm of the vector u.

The restricted isometry property holds for many pairs of bases including delta spikes and Fourier sinusoids. Interestingly, for a random, noise-like matrix Ψ (i.e., Gaussian random generated via randn in MATLAB®), with an overwhelmingly high probability, the matrix will be incoherent with any fixed basis, thus satisfying the restricted isometry property [57]. For signal-sensing applications, including radar imaging, this random sampling allows for accurate data compression at a rate much lower than Nyquist presented.

For the received radar signal in (5.1) or (5.2), the sparse target reflectivity function $s(t)$ or $s[n]$ can thus be estimated using the l_1-minimization problem and the random sensing matrix $\boldsymbol{\Psi}$ as

$$\hat{s} = \min_{s} \|s\|_1 \quad \text{subject to} \quad \|z - \boldsymbol{\Psi} Xs\| \leq \epsilon \ . \tag{5.6}$$

The use of compressive sensing for both radar imaging and target echo signal detection provides the possibility of significantly lowering the number of samples/measurements needed according to the Nyquist sampling criterion [60]. Unfortunately, practical implementations utilizing Gaussian and sub-Gaussian random variables require the use of M correlation channels among the hardware, allowing for time-domain signal correlations. For even slightly large values of M, this becomes infeasible.

Applications of compressive sensing for echo signal detection and image formation include work by Baraniuk [57] who demonstrated the capability to eliminate the matched filter in the receiver, as well as reducing the required A/D conversion bandwidth. Sampling rate reductions have additionally been shown for 1-D echo signal detection [61] and 2-D image formation [58, 62]. Shastry [58] proposed transmitting stochastic waveforms allowing for the reflected time-domain signal to be sampled at a lower rate while still retaining the properties necessary for the signal reflectivity to be accurately restored. Carin [62] demonstrated that with prior knowledge of the environment, using a two-way Green's function, the scattered fields can be inverted for image generation using a relatively small number of compressive sensing measurements. Gao [61] proposed the use of a waveform-matched dictionary, consisting of time-domain shifted versions of the transmitted signal, for 1-D target echo signal detection. The return signal clearly has a sparse representation using the waveform-matched basis allowing for a simplistic approach that can reduce the sampling rate while providing adequate target signal detection among clutter and noise. Chi [63] analyzed the performance degradations as a result of mismatches in the assumed basis for sparsity and the actual sparsity basis. The special care that must be taken to account for the mismatch was shown to be particularly important for the problem of image inversion common to radar and sonar systems.

5.2 ANGLE-DOPPLER-RANGE ESTIMATION

Early work with antenna arrays focused on directive radiation patterns. Known primarily as phased array radar systems, the relative phase of a signal to be transmitted was varied for each antenna, allowing for both reinforcement and suppression of the radiation pattern in certain directions. More recently, considerable attention has been paid to the exploitation of independent transmission and reception of signals at antenna arrays [64]. Whereas beamforming presumes a high-correlation between signals either transmitted or received by an array, multi-input multi-output (MIMO) radar systems utilizing separate signals at each antenna have been shown to have the ability to improve target resolution for both widely separated [65] and co-located antennas [66].

The introduction of a distributed MIMO radar system in 2008 [67] utilized a small scale network of randomly located nodes linked by a fusion center. Under the assumption that target presence was sparse in the angle-Doppler space, a compressive sensing based MIMO radar system was developed to extract target angle and Doppler (velocity) information. The result was not only superior resolution over a conventional pulsed radar system but the acquisition of target characteristics using far fewer samples than required by Nyquist.

Subsequent work has focused on improving target resolution, including range estimates, through the use of step-frequency radar (SFR). Recalling from Section 4.2 that the range resolution for a pulsed radar system is inversely proportional to the bandwidth of the transmitted pulse, wideband signals required for enhanced range resolution suffer not only from low signal-to-noise ratio (SNR) but carry additional computational burden due to the need for high-speed ADCs and processors [68]. SFR systems transmit several narrowband pulses at different frequencies. The frequency remains constant during each pulse resulting in a narrow instantaneous bandwidth. The range in transmitted frequencies allows for a large effective bandwidth, enhancing range resolution. Decoupled schemes using two separate pulse trains, one with a constant carrier frequency and the other that varies, have been introduced to reduce the complexity of jointly estimating angle, Doppler and range [69]. Compressive sensing was introduced to reduce the large number of pulses needing to be transmitted for an inverse discrete Fourier transform (IDFT) detector.

As a simple example, consider the early range and speed estimation approach presented by Shah [68]. Using co-located transmitters and receivers and N transmitted pulses, the target scene can be decomposed into an $M \times L$ range-speed plane with discretized range spaces of $[R_1, \ldots, R_M]$ and speed spaces of $[v_1, \ldots, v_L]$. Representing the target scene as a matrix S of size $M \times L$, the output of the phase detector for the reflected signal at distance R_m moving at speed v_l is [68]

$$y[k] = \sum_{m=1}^{M} \sum_{l=1}^{L} e^{i2\pi f_k \frac{2}{c}(R_m + v_l kT)} \cdot S(m, l) + w[k] , \tag{5.7}$$

where

$$S(m, l) = \begin{cases} \alpha & \text{reflectivity of target present at } (R_M, v_l) \\ 0 & \text{target is absent at } ((m-1)L + l)^{th} \text{ grid point} \end{cases} \tag{5.8}$$

and $w[k]$ represents zero-mean white noise. In matrix notation, this takes the familiar form

$$y = Ds + w , \tag{5.9}$$

where $s = [s_1, s_2, \ldots, s_{ML}] \in \mathbb{R}^{ML \times 1}$ is the rasterisation of the scene reflectivity matrix S. The basis matrix D consists of column vectors $\{d_i\}_{i=0}^{ML-1}$ of size $N \times 1$. The elements of the matrix can be shown to be

$$\varphi(k, (m-1)L + l) = e^{i2\pi(f_0 + k\Delta f)\frac{2}{c}(R_m + v_l kT)} , \tag{5.10}$$

where $k = 0, 1, \ldots, N-1$ and N again is the number of transmitted pulses. The column vectors of the basis matrix D correspond exactly to the phase detector outputs for all N pulses with a phase shift equivalent to that for a target located at the ith grid point. Assuming the measurement matrix Ψ in (5.6) is the identity matrix, simple convex optimization techniques such as the Dantzig selector can be used to recover the signal s with surprisingly high probability [17]. More importantly, since the matrix D consists of all possible range-speed combinations, the compressive sensing approach to range-speed estimation does not suffer from the shifting and spreading effects common for moving targets using the IDFT [68]. Compressive sensing techniques have also led to the use of non-identity measurement matrices to reduce data sampling rates and allow for the decoupling of range, speed and angle estimation [69].

5.3   IMAGE REGISTRATION (MATCHING) AND CHANGE DETECTION FOR SAR

Synthetic aperture radar (SAR) uses airborne or spaceborne mounted radar sensors to synthesize the large antenna aperture sizes needed for adequate cross range resolution. This technique is required in addition to the pulse compression technologies used to improve range resolution. Figure 5.1 shows the difference between SAR and optical imagery. Current SAR techniques produce near optical quality imagery with the added benefits of being able to be formed at long stand-off distances, under the cover of night and even in adverse weather conditions [28]. The obvious utilities of such an amazing technology are too large to enumerate, but we will focus on two topics that continue to see development throughout the literature. The automatic generation of scene height profiles, known as digital terrain map (DTM) generation, will be discussed in this section and the ability to automatically classify objects within the scene will be discussed in the ensuing section.

One benefit of SAR imagery is the coherent nature in which the images are formed. The process allows for coherent signal processing techniques such as interferometry [70] in which 3-D maps of the Earth's surface can be generated. The ability to do so is a result of the collection of the *complex* reflectivity of the illuminated scene. When the collection geometries for two separate SAR images are very similar, the two images can be *interfered* with each other so as to cancel the common reflectivities allowing for the inference of the scene geometries [70]. A typical scenario, known as one-pass interferometric SAR (IFSAR), utilizes two separate antennas on a single platform, differing only in depression angle geometry, to achieve extremely precise height estimation. The estimation precision is a direct consequence of the use of relatively short (on the order of centimeters) transmission wavelengths within the SAR sensing system.

An alternative approach to DTM generation that continues to see interest is the use of stereo radar image pairs. Since the electromagnetic reflectivity of all objects in a scene are laid over (projected) into a two-dimensional (range and cross range) imaging plane as a function of their height, different collection geometries produce different height-dependent placements, creating parallax between targets with height relief [71]. The potential for greater estimation accuracy increases with the parallax, or the angular difference between the SAR collection geometries. The benefit of the stereo approach is the known exact analytical solution for target heights in the scene based on the target disparity between image pairs.

Common to both approaches is the requirement for accurate feature correspondence between the image pairs. Image registration for natural imagery has always been a hot-bed of research, but the problem proves to be even more difficult for SAR imagery. In addition to the sometimes debilitating

Figure 5.1: Ku-band (15 GHz) SAR image (left) and optical image (right) of Albuquerque International Airport. (Image courtesy of Sandia National Labs, `http://www.sandia.gov/radar/imageryku.html`.)

multiplicative noise known as speckle, large translations, rotations, and image distortions are common for both imaging scenarios. For one-pass IFSAR, translations and rotations are mitigated by the use of multiple sensors aboard a single platform, but registration between image pairs is still an issue due to the height-dependent projection on the different imaging planes [28]. For stereo SAR, the large crossing angles required for greater parallax increase the effect of distortion and target layover as a result of the different collection geometries. An extreme example would be the different target signatures that can be expected when viewing a complex target from two completely opposite sides.

An example of a digital elevation map using IFSAR imagery for the capitol building in Washington, D.C. is shown in Figure 5.2. While the technique is amazingly accurate, the spatial distortions still prove difficult for registration and can severely impact the accuracy of the height estimates. Previous approaches for SAR image registration have relied upon both the coherent nature of the collection process, as well as the complex reflectivity (magnitude and phase) of each image pixel. These traditional translation-only complex correlation searches subdivide the image and find the translation-dependence required for maximal correlation between image pairs. Recent approaches [71] have used a shift-scale complex correlation approach to mitigate the effects of substantial height relief.

The problem of image registration, specifically for SAR imagery, was recently addressed by Nguyen and Tran [72]. Their approach to identifying the correspondence between local patches in the *reference* and *test* images was to create a dictionary using all overlapping patches of the reference image within a specified search region and find the best sparse approximation for each test image patch using a linear combination of the reference patches stored in the dictionary. Similar to the approach by Wright [8] used to account for image occlusion, an additional scaled identity matrix is appended to the dictionary to account for image noise.

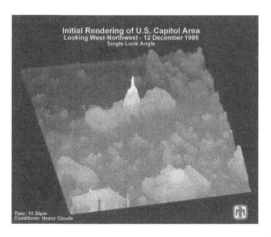

Figure 5.2: Digital elevation map for U.S. Capitol area generated using IFSAR imagery. (Image courtesy of Sandia National Labs, http://www.sandia.gov/radar/imageryku.html.)

Using image patches of size $M \times N$, a d-element dictionary $D \in \mathbb{R}^{MN \times d}$ and a test image patch $y \in \mathbb{R}^{MN \times 1}$, the problem of determining a sparse representation becomes

$$y = [D \; \mu I] \begin{bmatrix} \alpha_D \\ \alpha_I \end{bmatrix} = A\alpha \, , \tag{5.11}$$

where I is the $MN \times MN$ identity matrix, μ is a positive scalar (mean of reference search area), α_D are the coefficients associated with the dictionary elements and α_I are the coefficients associated with the identity matrix. Again, the addition of a scaled identity matrix helps to account for pixel irregularities in the approximation caused by image noise. The sparse solution for α can be computed using any of the techniques discussed in Section 2.3. The final approximation for the test image patch is given by

$$\hat{y} = D\widehat{\alpha_D} \, . \tag{5.12}$$

The approach does not determine the spatial displacement of the pixels between the two images, although it is easy to imagine a simple solution to determine such a measure needed for DTM generation. Instead, the authors considered the additional task of identifying changes that exist between the two images. Known as change detection, the second challenge relies on the suppression of the target signatures from both images. The authors pose a novel solution that includes the use of both the estimation accuracy and the degree of sparsity required for the estimation. The basic hypothesis being that when able to easily and accurately represent a test image patch using only a few non-zero coefficients, no change has occurred. By considering both the estimation accuracy and the degree of sparsity, a general measure of the amount of change observed can be estimated.

Change detection is only a single application once accurate image registration and pixel correspondence has been achieved. While extremely interesting, all applications can unfortunately not be considered. We refer the interested reader to the voluminous literature on stereo SAR [71], environmental monitoring [73], and moving target indication [74], among others.

5.4 AUTOMATIC TARGET CLASSIFICATION

As mentioned in Chapter 2, a byproduct of the computation of a compact and accurate signal within the sparse representation framework is the ability to infer further details about the signal itself. A significant amount of time is being devoted to the exploration of the ability for sparse representations to aid in object classification in areas including handwritten digits [75], facial recognition [8], and of course radar imagery [76].

A strong desire exists, particularly in military scenarios, to automatically detect and classify a wide variety of objects using high-resolution imagery. SAR imagery, with its multitude of benefits mentioned in the previous section, is a perfect candidate for object recognition. Early versions of automatic target recognition (ATR) systems utilizing SAR imagery have predominantly consisted of three stages [77]: an early detection stage that identifies local regions of interest using a constant false alarm rate (CFAR) detector, a one-class discrimination stage that aims to eliminate false-alarms while passing targets and the final classification stage responsible for classifying the remaining detections against a set of known target types.

In identifying regions of interest, the detection stage generates false alarms. No matter the classification system, subsequent stages are responsible for rejecting natural and man-made false alarms commonly referred to as *clutter*. While intuitive and easy to implement, template-matching based algorithms [78] correlate image detections with training templates for each target type. These templates are typically created at $5°$ increments in aspect or rotation angles of the target. The obvious crux of the approach is the intensive nature of correlating each detection with each template, particularly as more targets become known. Feature-based approaches, which use either computed image features or the training images directly, have been shown to be highly effective in target generalization and confuser rejection. Implementations such as the multilayer perceptron and support vector machine (SVM) have typically relied upon target pose estimation techniques which have proven to be particularly difficult in the presence of image speckle [79]. Recent studies have focused on finding a single discrimination/classification solution that is both efficient and accurate both in target classification and clutter rejection.

Building on both feature-based (training images used directly) and template matching (nearest neighbor approximation) algorithms, the use of sparse representations for SAR target classification was recently explored using two separate image processing techniques. In the next two sections, we present our work on target classification in SAR imagery using sparse representations. These algorithms have been shown to be effective at classifying targets in SAR imagery, however, they continue to be refined in hopes of developing the next state-of-the-art SAR target classification scheme.

5.4.1   SPARSE REPRESENTATION FOR TARGET CLASSIFICATION

Emulating the approach presented by Wright [8] for facial recognition, we recently presented the extension of sparse representation based target classification for SAR imagery [76]. The approach was motivated from the perspective of manifold approximation, wherein test images for a given class were assumed to be sampled from an underlying manifold composed of training images of the same type. Under the assumption that the test images do not lie far from the class manifold, a reasonable linear approximation based on the training images would then be sufficient to represent the test image. Given two manifolds, $(\mathcal{M}_1, \mathcal{M}_2)$ as shown in Figure 5.3, the linear approximations for test image x onto the jth, $j = 1, 2$ class manifolds are given by

$$\hat{x}_J = \sum_{k \in \Delta_x^j} \alpha_{k,j} d_{k,j} \, , \tag{5.13}$$

where Δ_k^j is the set containing the indices k of the training vectors from class j so that $d_{k,j}$ is the kth

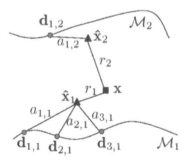

Figure 5.3: Classification of test data (x) using local linear projections ($\widehat{x}_1, \widehat{x}_2$) and residuals ($r_1, r_2$) on the manifolds ($\mathcal{M}_1, \mathcal{M}_2$).

dictionary element for class j. The training images for all known classes are combined into a single dictionary, D, so that the approximation coefficient vector α can be computed by solving $x = D\alpha$ using one of the l_1-norm solution methods discussed in Section 2.3. Classification decisions are made by selecting the target class j that minimizes the projected image residual, i.e.,

$$c = \arg\min_j \|\boldsymbol{r}_j\|_2 = \arg\min_j \|\boldsymbol{x} - \hat{\boldsymbol{x}}_j\| . \qquad (5.14)$$

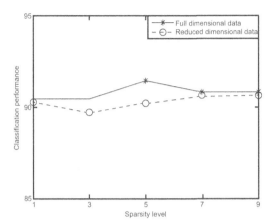

Figure 5.4: Classification performance in full and reduced dimensional cases [76]. Target images collected at a 17° depression angle were used for training and 15° depression angles images were used for testing.

Results for 3 targets (T72, BMP2, BTR70) from the public Moving and Stationary Target Acquisition and Recognition (MSTAR) database [54] are shown in Figure 5.4. The solid line indicates classification accuracy using the full 128×128 power-domain imagery while the dashed line indicates performance results when using random projections to reduce the image dimensionality by almost 90% to 1792 pixels. Results above 90% P_{cc} are virtually equivalent to that for recent work done using machine learning algorithms. More importantly, however, is the lack of performance degradation when using random projections for dimensionality reduction in conjunction with the sparse learning architecture.

5.4.2 SPARSE REPRESENTATION-BASED SPATIAL PYRAMIDS

We have recently been working to extend the sparse representation framework [80] to work in conjunction with an emerging and highly useful classification/recognition architecture known as spatial pyramid matching [81]. Initially designed to address the issue of scene categorization, the spatial pyramid matching scheme was shown to be useful for recognition of objects within the scene itself. The ability to do so lies primarily in the approximate geometric correspondence that is established when systematically aggregating image features over fixed sub-regions in the image. The integration of the sparse representation framework into the feature pooling portion of the algorithm improves the spatial pyramid framework by identifying only the strongest dictionary elements within each sub-region.

Consider the generation of a three-level spatial pyramid as shown in Figure 5.5. Using a generic dictionary, $\boldsymbol{D} \in \mathbb{R}^{d \times F}$, where d is the number of elements in the dictionary and F is the length of an arbitrary image feature vector, local image features such as scale-invariant feature transforms (SIFTs) or even basic FFTs can be encoded to provide a feature descriptor for *interesting* points in the image. Currently, dictionary generation is accomplished using clustering of all feature descriptors from the training imagery, although more sophisticated dictionary generation algorithms continue to be explored.

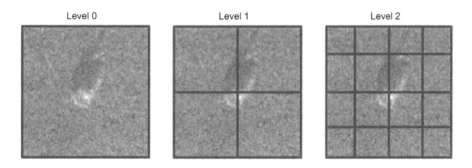

Figure 5.5: Generation of a three-level spatial pyramid.

For each sub-region at each level L sparse approximations are used to encode the local image features using the dictionary $\boldsymbol{D}$. Given f_l feature vectors for a sub-region, the sparse representation framework generates a coefficient matrix $\boldsymbol{C} \in \mathbb{R}^{d \times f_l}$ where each element c_{ij} corresponds to the representation coefficient for the ith dictionary element for feature vector j. Contrast this to the vector quantization approach presented by Lezabnik, where each column of $\boldsymbol{C}$ contains a single unit element indicating the most similar (l_2-norm) dictionary atom for that feature vector. A single feature descriptor $\boldsymbol{f}$ for each sub-region is then generated from $\boldsymbol{C}$ using either average (histogram binning) or max pooling. Mathematically, these are equivalent to

$$\text{Average Pooling:} \quad \boldsymbol{f} = \sum_j c_{ij}$$
$$\text{Max Pooling:} \quad \boldsymbol{f} = \max_j c_{ij} \, .$$

Average pooling identifies the total number of dictionary elements present in each sub-region while max pooling identifies the maximum component for each dictionary element (if present at all) in the sub-region.

The final feature vector for the entire image is the concatenation of each sub-region feature vector f. The length $d\frac{1}{3}(4^L - 1)$ vector is itself very sparse, a property which we seek to exploit in further research. For the time being, the *long* feature vector is used as the input to a one-versus-one linear SVM, whose raw output score for each class will be used to make the classification decision. Preliminary results using SIFT features and 16×16 FFT patch features for the same three targets from the MSTAR data set from Section 5.4.1 are shown in Table 5.1. The use of sparse representations and max pooling clearly improve classification performance for SAR imagery. The nearly 10% increase in classification performance when using FFT features rather than SIFT features indicates the potential for significant improvement when more advanced image features, tuned specifically for SAR imagery, are used.

Table 5.1: MSTAR classification results using spatial pyramid matching								
	M = 100			M = 200			M = 400		
	L = 1	L = 5	L = 10	L = 1	L = 5	L = 10	L = 1	L = 5	L = 10
SIFT w/Histograms		73.19%			77.14%			82.78%	
SIFT w/SR	77.73%	78.32%	77.14%	80.29%	79.27%	78.39%	81.54%	79.27%	79.27%
FFT w/Histograms		84.91%			87.33%			87.91%	
FFT w/SR	**88.94%**	87.25%	80.88%	**90.26%**	86.81%	81.69%	**88.57%**	88.13%	82.56%

Moving forward, it is expected that the utility of the approach will be exemplified by its ability to handle target occlusion. The ability to maintain geometric correspondence while *segmenting* the target image into parts could prove to be extremely useful when the target is partially hidden, either intentionally or unintentionally. In addition, if indeed geometric correspondence between target features is maintained, it is expected that the algorithm will prove to be robust to target translations. Translation and occlusion independence have proven to be difficult tasks for even the most advanced SAR detection algorithms throughout the literature. An algorithm that is able to successfully handle these and other unexpected non-benign operating conditions could very well be the next state-of-the-art SAR automatic target recognition system.

APPENDIX A

Code Sample

A.1 NON-UNIFORM SAMPLING AND SIGNAL RECONSTRUCTION CODE

```
%nonUniform_Sampling.m - Non-Uniform Sampling
clear; clc; close all
f1 = 1e3; f2 = 2e3; f3 = 4e3;   %Hz
fs = 10e3;                       %Hz
T = 1/fs;                        %Period in seconds

%Number of samples
N = 1028;                        %For .1s of data
t = [0:N-1]'*T;

%Full-length and sampled signals
x = sin(2*pi*f1*t) + sin(2*pi*f2*t) + sin(2*pi*f3*t);

%Generate random samples
M = 151;                         %Approximately 10% of total signal
k = randperm(N);
m = k(1:M);
b = x(sort(m));

%Plot time-domain signal
figure;subplot(3,2,1);plot(t,x,'r-',t(m),b,'bo');
xlabel('Time (s)');ylabel('x(t)');title('Sum of sinusoids');

%Plot (sparse) Fourier domain signal
X = fft(x);
subplot(3,2,2);plot(linspace(0,fs,length(X)),abs(X));
xlabel('Frequency (Hz)');ylabel('|X(f)|');

%Generate linear transform matrix
PHI = fft(eye(N));
S = zeros(M,N);
S(sub2ind(size(S),1:M,sort(m))) = 1;
A = S*(1/N)*PHI';
```

```
%Naive l2 minimization solution
l2 = pinv(A)*b;
subplot(3,2,4);plot(linspace(0,fs,length(X)),abs(l2));
xlabel('Frequency (Hz)');ylabel('|X_l_2(f)|');

%Reconstruct signal
xl2 = (1/N)*real(PHI'*l2);
subplot(3,2,3);plot(t,xl2);
xlabel('Time (s)');ylabel('x_l_2(t)');

%Approximate l1 solution using OMP
l1 = OMPnorm(A,b,floor(M/4),0);
subplot(3,2,6);plot(linspace(0,fs,length(X)),abs(l1));
xlabel('Frequency (Hz)');ylabel('|X_l_1(f)|');


%Reconstruct signal
xl1 = (1/N)*real(PHI'*l1);
subplot(3,2,5);plot(t,xl1);
xlabel('Time (s)');ylabel('x_l_1(t)');
xlabel('x_p');ylabel('y_p');title('PCA Transformation w/2
Eigenvectors');
```

A1: 1-D Signal Reconstruction Analysis

A.2   LONG-SHEPP PHANTOM TEST IMAGE RECONSTRUCTION CODE

```matlab
%*********************************************************************
% DISCLAIMER: This code has been modified from the original
% tveq_phantom.m script supplied with l1-MAGIC package.  It has been
% updated to generate a single figure containing the test image,
% Fourier sampling mask, minimum energy solution, and total
% variation solution.
%*********************************************************************
clear; clc; close all;
figure;

% Phantom
n = 256;                        %Size of image
N = n*n;                        %Number of pixels
X = phantom(n);
x = X(:);                       %Rasterize image

% Number of radial lines in the Fourier domain
L = 22;

% Fourier samples we are given
[M,Mh,mh,mhi] = LineMask(L,n);
OMEGA = mhi;
A = @(z) A_fhp(z, OMEGA);
At = @(z) At_fhp(z, OMEGA, n);

% measurements
y = A(x);

% min l2 reconstruction (backprojection)
xbp = At(y);
Xbp = reshape(xbp,n,n);

% recovery
tvI = sum(sum(sqrt([diff(X,1,2) zeros(n,1)].^2 ...
    [diff(X,1,1); zeros(1,n)].^2 )));
disp(sprintf('Original TV = %8.3f', tvI));
xp = tveq_logbarrier(xbp, A, At, y, 1e-1, 2, 1e-8, 600);
Xtv = reshape(xp, n, n);
```

```
%Plot everything
subplot(2,2,1);imagesc(X);colormap gray;axis image;axis off;
xlabel('(a)');
subplot(2,2,2);imagesc(fftshift(M));colormap gray;axis image;axis
off;xlabel('(b)');
subplot(2,2,3);imagesc(Xbp);colormap gray;axis image;axis off;


xlabel('(c)');
subplot(2,2,4);imagesc(Xtv);colormap gray;axis image;axis off;
xlabel('(d)');
```

A2: 2-D Phantom Image Reconstruction Using Total Variation

A.3   SIGNAL BANDWIDTH CODE

```
%Ex - 4.1 - Pulse radar signal and bandwidth examples
clear; clc; close all

%Operating parameters
n = 1;                     %Length of signal (in seconds)
DC = .05;                  %Duty cycle
tau_b = DC*n;              %10% Duty Cycle
f = 500; fs = 5000;
t = 0:1/fs:n;

% Generate envelope functions
cenv          = .5*(1+cos(2*pi*(t-tau_b/2)/tau_b));
cenv(t>tau_b) = 0;
sqp           = ones(1,length(cenv));
sqp(t>tau_b)  = 0;

%Part (a) - CW Pulse
x_ncenv    = cenv.*sin(2*pi*f*t);
x_nsqp     = sqp.*sin(2*pi*f*t);
figure;plot(t,x_ncenv);xlim([0 tau_b]);

%Part (b) - Chirp Pulse
x_c        = chirp(0:1/fs:tau_b,f-.1*f,tau_b,f+.1*f);
x_c        = [x_c zeros(1,length(t)-length(x_c))];
x_ccenv    = cenv.*x_c;
x_csqp     = sqp.*x_c;
figure;plot(t,x_csqp);xlim([0 tau_b]);

% Part (c) - Frequency Spectrum for Rectangular Pulse
fn = linspace(-fs/2,fs/2,length(t));
figure;
plot(fn,db(abs(fftshift(fft(x_nsqp)))),'-k');hold on;
plot(fn,db(abs(fftshift(fft(x_csqp)))),'-r');
xlim([f-.5*f f+.5*f]);legend('CW Pulse','Chirp');
xlabel('f');ylabel('dB(X_f)');
```

```
% Part (d) - Frequency Spectrum for Raised Cosine Pulse
fn = linspace(-fs/2,fs/2,length(t));
figure;
plot(fn,db(abs(fftshift(fft(x_ncenv)))),'-k');hold on;
plot(fn,db(abs(fftshift(fft(x_ccenv)))),'-r');
xlim([f-.5*f f+.5*f]);legend('CW Pulse','Chirp');
xlabel('f');ylabel('dB(X_f)');
```
A3:Signal bandwidhts for CW and compressed waveforms

Bibliography

[1] H. Hertz, *Electric Waves*. New York: Dover Publications, 1962. (Republication of the work first published in 1983 by Macmillan and Company.) Cited on page(s) 1

[2] M. A. Richards, *Fundamentals of Radar Signal Processing*. New York: McGraw-Hill, 2005. Cited on page(s) 1, 2, 33, 36, 37, 39, 41, 42

[3] M. Elad, *Sparse and Redundant Representations*. New York: Springer, 2010. DOI: 10.1007/978-1-4419-7011-4 Cited on page(s) 1, 10, 12

[4] M. Skolnik, *Introduction to Radar Systems*, 3rd ed. New York: McGraw-Hill, 2007. Cited on page(s) 1, 2, 3, 33, 35, 36, 45

[5] S. G. Marconi, "Radio Telegraphy," in *Proc. IRE*, vol. 10, 1992, p. 237. Cited on page(s) 1

[6] J. Rissanen, "Modeling by shortest data description," *Automatica*, vol. 14, pp. 465–471, 1978. DOI: 10.1016/0005-1098(78)90005-5 Cited on page(s) 7

[7] J. B. Tenenbaum, V. de Silva and J. C. Langford, "A global geometric framework for nonlinear dimensionality reduction," *Science*, vol. 290, no. 5500, pp. 2319–2323, 2000. DOI: 10.1126/science.290.5500.2319 Cited on page(s) 7, 25, 26, 27

[8] J. Wright et. al., "Robust face recognition via sparse representation," *IEEE Transactions on Pattern Analysis and Machine Intelligence*, vol. 31, no. 2, pp. 210–227, 2009. DOI: 10.1109/TPAMI.2008.79 Cited on page(s) 7, 8, 51, 52, 53

[9] E. Candès, J. Romberg and T. Tao, "Stable signal recovery from incomplete and inaccurate measurements," *Communications on Pure and Applied Mathematics*, vol. 59, no. 8, pp. 1207–1223, 2005. DOI: 10.1002/cpa.20124 Cited on page(s) 7, 18

[10] D. L. Donoho, "For most large undeterdetermined systems of linear equations the minimal l_1-norm solution is also the sparsest solution," *Communications on Pure and Applied Mathematics*, vol. 59, no. 6, pp. 797–829, 2006. DOI: 10.1002/cpa.20132 Cited on page(s) 7, 9

[11] E. Candes, J. Romberg and T. Tao, "Robust uncertainty principles: Exact signal reconstruction from highly incomplete frequency information," *IEEE Transaction on Information Theory*, vol. 52, no. 2, pp. 489–509, February 2006. DOI: 10.1109/TIT.2005.862083 Cited on page(s) 8

[12] G. Strang, *Introduction to Linear Algebra*, 4th ed. Wesseley, MA: Wesseley-Cambridge Press, 2009. Cited on page(s) 8

[13] I. Daubechies, "Time-frequency localization operators: a geometric phase space approach," *IEEE Transaction on Information Theory*, vol. 34, pp. 605–612, 1988. DOI: 10.1109/18.9761 Cited on page(s) 8

[14] M. R. Garey and D. S. Johnson, *Computers and Intractability: A Guide to Theory of NP-completeness*. San Francisco: W. H. Freeman and Company, 1979. Cited on page(s) 9

[15] E. Amaldi and V. Kann, "On the approximability of minimizing nonzero variables or unsatisfied relations in linear systems," *Theoretical Computer Science*, vol. 209, pp. 237–260, 1998. DOI: 10.1016/S0304-3975(97)00115-1 Cited on page(s) 9

[16] S. Chen, D. Donoho, and M. Saunders, "Atomic decomposition by basis pursuit," *SIAM Review*, vol. 43, no. 1, pp. 129–159, 2001. DOI: 10.1137/S003614450037906X Cited on page(s) 9, 12

[17] E. Candes and T. Tao, "The Dantzig selector: Statistical estimation when p is much larger than n," *Annals of Stastistics*, vol. 35, pp. 2313–2351, 2007. DOI: 10.1214/009053606000001523 Cited on page(s) 12, 48

[18] S. Mallat and Z. Zhang, "Matching pursuits with time-frequency dictionaries," *IEEE Transactions on Signal Processing*, vol. 41, no. 12, pp. 3397–3415, 1993. DOI: 10.1109/78.258082 Cited on page(s) 12

[19] Y. C. Pati, R. Rezaifar and P. S. Krishnaprasad, "Orthogonal matching pursuit: recursive function approximation with applications to wavelet decomposition," in *27th Asilomar Conf. on Signals, Systems and Comput.*, Nov. 1993. DOI: 10.1109/ACSSC.1993.342465 Cited on page(s) 12

[20] M. E. Davies and T. Blumesath, "Faster and greedier: algorithms for sparse reconstruction of large datasets," in *Proceedings of ISCCSP 2008*, 2008, pp. 774–779. DOI: 10.1109/ISCCSP.2008.4537327 Cited on page(s) 12, 13, 16

[21] J. H. Friedman and W. Steutzle, "Projection pursuit regression," *American Statistics Association*, vol. 76, pp. 817–823, 1981. DOI: 10.1080/01621459.1981.10477729 Cited on page(s) 13

[22] T. Blumensath and M. E. Davies, "Gradient pursuits," *IEEE Transactions on Signal Processing*, vol. 56, no. 6, pp. 2370–2382, 2008. DOI: 10.1109/TSP.2007.916124 Cited on page(s) 15

[23] J. A. Tropp and A. C. Gilbert, "Signal recovery from random measurements via orthogonal matching pursuit," *IEEE Transactions on Information Theory*, vol. 53, no. 12, pp. 4655–4666, December 2007. DOI: 10.1109/TIT.2007.909108 Cited on page(s) 15

[24] E. Candès, "Compressive Sampling," in *International Congress of Mathematicians*, Madrid, Spain, 2006, pp. 1433–1452. Cited on page(s) 15

[25] E. Candès. (2005, October) l_1 Magic. [Online]. `http://www.acm.caltech.edu/l1magic/` Cited on page(s) 17, 19

[26] C. Moler. (2010) 'Magic' Reconstruction: Compressed Sensing. [Online]. `http://www.mathworks.com/company/newsletters/articles/clevescorner-compressed-sensing.html?issue=nn2010` Cited on page(s) 17

[27] A. H. Delaney and Y. Bresler, "A fast and accurate iterative reconstruction algorithm," *IEEE Transactions on Image Processing*, vol. 5, pp. 740–753, 1996. DOI: 10.1109/83.495957 Cited on page(s) 17

[28] C. V. Jakowatz, D. E., Eichel, P. H. Wahl, D. C. Ghiglia, and P. A. Thompson, *Spotlight-Mode Synthetic Aperture Radar: A Signal Processing Approach*. New York: Springer Science + Business Media, 1996. DOI: 10.1007/978-1-4613-1333-5 Cited on page(s) 17, 34, 36, 37, 41, 49, 50

[29] E. J. Candès, J. Romberg and T. Tao, "Robust uncertainty principles: exact signal reconstruction from highly incomplete frequency information," *IEEE Transactions on Information Theory*, vol. 52, pp. 489–509, 2006. DOI: 10.1109/TIT.2005.862083 Cited on page(s) 18

[30] Z. Zhang and H. Zha, Local linear smoothing for nonlinear manifold learning, 2003, Technical Report, Zhejjang University. Cited on page(s) 21

[31] H. Hotelling, "Analysis of a complex of statistical variables into principal components," *Journal of Educational Psychology*, vol. 24, pp. 417–441, 1933. DOI: 10.1037/h0070888 Cited on page(s) 21

[32] I. T. Jollife, *Principal Component Analysis*. New York: Springer-Verlag, 2002. Cited on page(s) 21

[33] R. A. Fisher, "The use of multiple measurements in taxonomic problems," *Annals of Eugenics*, vol. 7, pp. 179–188, 1936. DOI: 10.1111/j.1469-1809.1936.tb02137.x Cited on page(s) 21

[34] L. J. P. Van Der Maaten, E. O., Van Den Herik and H. J. Postma, "Dimensionality reduction: A comparative overview," Submitted for publication to Elsevier, 2007. Cited on page(s) 21, 22, 23, 25, 28

[35] Y. Bengio and M. Monperrus, "Non-local manifold tangent learning," *Advances in Neural Information Processing Systems*, vol. 17, pp. 129–136, 2005. Cited on page(s) 21

[36] C. M. Bishop, *Pattern Recognition and Machine Learning*. New York: Springer Science + Business Media, 2006. Cited on page(s) 22, 23, 25

[37] K. Pearson, "On lines and planes of closest fit to systems of points in space," *The London, Edinburgh and Dublin Philosophical Magazine and Journal of Science, Sixth Series*, vol. 2, pp. 559–572, 1901. Cited on page(s) 22

[38] M. G. Partridge and R. A. Calvo, "Fast dimensionality reduction and simple PCA," *Intelligent Data Analysis*, vol. 2, no. 1, pp. 203–214, 1998. DOI: 10.1016/S1088-467X(98)00024-9 Cited on page(s) 23

[39] S. Roweis, "EM Algorithms for PCA and SPCA," in *Advances in Neural Information Processing Systems*, 1998, pp. 626–632. Cited on page(s) 23

[40] J. Yin, D. Hu and Z. Zhou, "Noisy manifold learning using neighborhood smoothing embedding," *Pattern Recognition Letters*, vol. 29, no. 11, pp. 1613–1620, 2008. DOI: 10.1016/j.patrec.2008.04.002 Cited on page(s) 25

[41] K. Q. Weinberger and L. K. Saul, "An introduction to nonlinear dimensionality reduction maximum variance unfolding," in *Proceedings of the 21st National Conference on Artificial Intelligence*, 2006. Cited on page(s) 27

[42] S. Lagon and A. B. Lee, "Diffusion maps and coarse-graining: A unified framework for dimensionality reduction, graph partitioning and data set parameterization," *IEEE Trans. Pattern Analysis and Machine Intelligence*, vol. 28, pp. 1393–1403, 2006. DOI: 10.1109/TPAMI.2006.184 Cited on page(s) 27

[43] B Scholkopf, A. J. Smola and K. R. Muller, "Nonlinear componenet analysis as a kernel eigenvalue problem," *Neural Computation*, vol. 10, no. 5, pp. 1299–1319, 1998. DOI: 10.1162/089976698300017467 Cited on page(s) 27

[44] S. T. Roweis and L. K. Saul, "Nonlinear dimensionality reduction by locally linear embedding," *Science*, vol. 290, pp. 2323–2326, 2000. DOI: 10.1126/science.290.5500.2323 Cited on page(s) 27, 28

[45] T. F. Cox and M. A. Cox, *Multidimensional Scaling*. London: Chapman and Hall, 2001. Cited on page(s) 27

[46] M. Belkin and P. Niyogi, "Laplacian eigenmaps and spectral techniques for embedding and clustering," *Advances in Neural Information Processing Systems*, vol. 14, pp. 585–591, 2002. Cited on page(s) 28

[47] D. L. Donoho and C. Grimes, "Hessian eigenmaps: New locally linear embedding techniques for high-dimensional data," in *Proceedings of the National Academy of Sciences*, vol. 102, 2005, pp. 7426–7431. DOI: 10.1073/pnas.1031596100 Cited on page(s) 28

[48] Z. Zhang and H. Zha, "Principal manifolds and nonlinear dimension reduction via local tangent space alignment," *SIAM Journal of Scientific Computing*, vol. 26, no. 1, pp. 313–338, 2004. DOI: 10.1137/S1064827502419154 Cited on page(s) 30

[49] Y. W. Teh and S. T. Roweis, "Automatic alignment of local representations," in *Advances in Neural Information Processing Systems*, vol. 15, 2003, pp. 841–848. Cited on page(s) 30

[50] M. Brand, "Charting a manifold," in *Advances in Neural Information Processing Systems*, vol. 15, 2002, pp. 985–992. Cited on page(s) 30

[51] G. H. Golub and C. F. van Loan, *Matrix Computations*. Oxford, UK: North Oxford Academic, 1983. Cited on page(s) 30

[52] S. Dasgupta and A. Gupta, "An elementary proof of the Johnson-Lindenstrauss lemma," U. C. Berkeley, Technical Report 99–006 Mar. 1999. Cited on page(s) 30

[53] E. Bingham and H. Mannila, "Random projection in dimensionality reduction: applications to image and text data," in *Knolwedge Discovery and Data Mining*, 2001, pp. 245–250. DOI: 10.1145/502512.502546 Cited on page(s) 31, 32

[54] E. R. Keydel, "MSTAR extended operating conditions," in *Proceedings of SPIE*, vol. 2757, 1996, pp. 228–242. DOI: 10.1117/12.242059 Cited on page(s) 32, 54

[55] D. E. Dudgeon and R. M. Mersereau, *Multidimensional Digital Signal Processing*. Englewood Cliffs, NJ: Prentice Hall, 1984. Cited on page(s) 41

[56] S. M. Kay, *Fundamentals of Statistical Signal Processing: Detection Theory*, 2nd ed. Upper Saddle River, NJ: Prentice Hall, 1993. Cited on page(s) 42

[57] R. Baraniuk and P. Steeghs, "Compressive radar imaging," in *IEEE Radar Conference*, Waltham, MA, Apr. 2007, pp. 128–133. DOI: 10.1109/RADAR.2007.374203 Cited on page(s) 45, 46, 47

[58] M. C. Shastry, R. M. Narayanan and M. Rangaswamy, "Compressive radar imaging using white stochastic waveforms," in *Proceedings of the 5th IEEE International Waveform Diversity and Design*, Aug. 2010, pp. 90–94. DOI: 10.1109/WDD.2010.5592367 Cited on page(s) 46, 47

[59] D. Donoho, "Compressed sensing," *IEEE Transactions on Information Theory*, vol. 52, no. 4, pp. 1289–1306, April 2006. DOI: 10.1109/TIT.2006.871582 Cited on page(s) 46

[60] H. Nyquist, "Certain topics in telegraph transmission theory," *Proceedings of the IEEE*, vol. 90, no. 2, pp. 280–305, Feb. 2002. DOI: 10.1109/5.989875 Cited on page(s) 47

[61] D. Gao, D. Liu, Y. Feng, Q. An and F. Yu, "Radar echo signal detection with sparse representations," in *Proceedings of the 2nd International Conference on Signal Processing Systems (ICSPS)*, July 2010, pp. 495–498. DOI: 10.1109/ICSPS.2010.5555846 Cited on page(s) 47

[62] L. Carin, D. Liu and B. Guo, "In situ compressive sensing multi-static scattering: Imaging and the restricted isometry property," preprint, 2008. Cited on page(s) 47

[63] Y. Chi, L. Scharf, A. Pezeshki and R. A. Calderbank, "Sensitivity to basis mismatch in compressed sensing," *IEEE Transactions on Signal Processing*, vol. 59, no. 5, pp. 2182–2195, May 2011. DOI: 10.1109/TSP.2011.2112650 Cited on page(s) 47

[64] E. Fishler et al., "MIMO radar: An idea whose time has come," in *Proc. IEEE Radar Conf*, Philadelphia, PA, Apr. 2004, pp. 71–78. DOI: 10.1109/NRC.2004.1316398 Cited on page(s) 47

[65] A. M. Haimovich, R. S. Blum and L. J. Cimini, "MIMO radar with widely separated antennas," *IEEE Signal Processing Magazine*, vol. 25, no. 1, pp. 116–129, 2008. DOI: 10.1109/MSP.2008.4408448 Cited on page(s) 47

[66] P. Stoica and J. Li, "MIMO radar with colocated antennas," *IEEE Signal Processing Magazine*, vol. 24, no. 5, pp. 106–114, 2007. DOI: 10.1109/MSP.2007.904812 Cited on page(s) 47

[67] A. P. Petropulu, Y Yu and H. V. Poor, "Distributed MIMO radar using compressive sampling," in *Proc. 42nd Asilmoar Conf. Signals, Syst. Comput.*, Pacific Grove, CA, Nov. 2008, pp. 203–207. Cited on page(s) 47

[68] S. Shah, Y. Yu and A. P. Petropulu, "Step-frequency radar with compressive sampling (SFR-CS)," in *Proc. ICASSP 2010*, Dallas, TX, Mar. 2010, pp. 1686–1689. DOI: 10.1109/ICASSP.2010.5495497 Cited on page(s) 48, 49

[69] Y. Yu, A. P. Petropulu and H. V. Poor, "Reduced complexity angle-Doppler-range estimation for MIMO radar that employs compressive sensing," in *Proceedings of the Forty-Third Asilomar Conference on Signals, Systems and Computers*, Nov. 2009, pp. 1196–1200. DOI: 10.1109/ACSSC.2009.5469995 Cited on page(s) 48, 49

[70] P. A. Rosen, "Synthetic aperture radar interferometry," *Proceedings of the IEEE*, vol. 88, no. 3, pp. 333–382, March 2000. DOI: 10.1109/5.838084 Cited on page(s) 49

[71] D. A. Yocky and C. V. Jakowatz, "Shift-scale complex correlation for wide-angle coherent cross-track SAR stereo processing," *IEEE Transactions on Geoscience and Remote Sensing*, vol. 45, no. 3, pp. 576–583, March 2007. DOI: 10.1109/TGRS.2006.886193 Cited on page(s) 49, 50, 52

[72] L. H. Nguyen and T. D. Tran, "A sparsity-driven joint image registration and change detection technique for SAR imagery," in *IEEE International Conference on Acoustics, Speech and Signal*

Processing, Mar. 2010, pp. 2798–2801. DOI: 10.1109/ICASSP.2010.5496197 Cited on page(s) 51

[73] C. T. Wang et al., "Disaster monitoring and environmental alert in Taiwan by repeat-pass space-borne SAR," in *International Geoscience and Remote Sensing Symposium*, Jul. 2007, pp. 2628–2631. DOI: 10.1109/IGARSS.2007.4423384 Cited on page(s) 52

[74] X. Wang, Y. Liu and Y. Huang, "The application of image registration based on genetic algorithm with real data," in *2nd Asian-Pacific Conference on Synthetic Aperture Radar*, Oct. 2009, pp. 844–847. DOI: 10.1109/APSAR.2009.5374187 Cited on page(s) 52

[75] K. Huang and S. Aviyente, "Sparse representation for signal classification," in *Advances in Neural Information Processing Systems*, 2006, pp. 609–617. Cited on page(s) 52

[76] J. Thiagarajan, K. Ramamurthy, P. Knee and A. Spanias, "Sparse representations for automatic target classification in SAR images," in *4th International Symposium on Communications, Control and Signal Processing (ISCCSP)*, Mar. 2010, pp. 1–4. DOI: 10.1109/ISCCSP.2010.5463416 Cited on page(s) 52, 53, 54

[77] D. E. Kreithen, S. D. Halversen and G. J. Owirka, "Discriminating targets from clutter," *Lincoln Laboratory Journal*, vol. 6, no. 1, pp. 25–52, 1993. Cited on page(s) 52

[78] G. J. Owirka, S. M. Verbout and L. M. Novak, "Template-based SAR ATR performance using different image enhancement techniques," in *Proceedings of SPIE*, vol. 3721, 1999, pp. 302–319. DOI: 10.1117/12.357648 Cited on page(s) 52

[79] Q. Zhao et al., "Support vector machines for SAR automatic target recognition," *IEEE Transactions on Aerospace and Electronic Systems*, vol. 37, no. 2, pp. 643–653, 2001. DOI: 10.1109/7.937475 Cited on page(s) 52

[80] P. Knee, J. Thiagarajan, K. Ramamurthy and A. Spanias, "SAR target classification using sparse representations and spatial pyramids," in *IEEE Internation Radar Conference*, Kansas City, MO, 2011. DOI: 10.1109/RADAR.2011.5960546 Cited on page(s) 54

[81] S. Lazebnik, C. Schmid and J. Ponce, "Beyond bags of features: Spatial pyramid matching for recognizing natural scene categories," in *IEEE Computer Society Conference on Computer Vision and Pattern Recognition (CVPR)*, vol. 2, 2006, pp. 2169–2178. DOI: 10.1109/CVPR.2006.68 Cited on page(s) 54

Author's Biography

PETER A. KNEE

Peter A. Knee received a B.S. (with honors) in electrical engineering from the University of New Mexico, Albuquerque, New Mexico, in 2006, and an M.S. degree in electrical engineering from Arizona State University in 2010. While at Arizona State University, his research included the analysis of high-dimensional Synthetic Aperture Radar (SAR) imagery for use with Automatic Target Recognition (ATR) systems as well as dictionary learning and data classification using sparse representations. He is currently an employee at Sandia National Laboratories in Albuquerque, New Mexico, focusing on SAR image analysis and software defined radios.

Printed in the United States
by Baker & Taylor Publisher Services